T-Labs Series in Telecommunication Services

Series editors

Sebastian Möller, Berlin, Germany
Axel Küpper, Berlin, Germany
Alexander Raake, Berlin, Germany

More information about this series at http://www.springer.com/series/10013

Patrick Stewin

Detecting Peripheral-based
Attacks on the Host Memory

 Springer

Patrick Stewin
Technische Universität Berlin
Berlin
Germany

ISSN 2192-2810   ISSN 2192-2829   (electronic)
T-Labs Series in Telecommunication Services
ISBN 978-3-319-13514-4   ISBN 978-3-319-13515-1   (eBook)
DOI 10.1007/978-3-319-13515-1

Library of Congress Control Number: 2014955796

Springer Cham Heidelberg New York Dordrecht London

Printed on acid-free paper

Springer International Publishing AG Switzerland is part of Springer Science+Business Media
(www.springer.com)

To Gesche

Acknowledgments

First of all, I would especially like to thank my advisor Jean-Pierre Seifert. I am not only grateful for many useful discussions and the excellent research environment, but also for leaving me free to select my own thesis topic. His infections, encouragement, motivation, and inspiration were always greatly appreciated. Thanks to him I always believed in my research and my thesis. Next, I would like to extend my sincerest thanks to my colleagues and friends from the Chair for Security in Telecommunications (SecT) at TU Berlin. Special thanks go to Nico Golde and Dmitry Nedospasov (the Ph.D. team!) as well as Iurii Bystrov, Kévin Redon, Ravi Borgaonkar, and Collin Mulliner. Without the Ph.D. team I would still be working on my thesis. Specifically, I thank Collin for his advice in all areas. Without Iurii the Intel AMT/ME-related projects would not have been such a great success.

I also would like to thank the Communication and Operating Systems (KBS) research group as well as the Workgroup for Computer Security (AGRS) at TU Berlin for many useful comments as well as for their helpful suggestions to prepare conference talks. Additionally, I would like to extend my thanks to Dirk Kuhlmann and Chris Dalton from the Cloud and Security Lab (HP Labs Bristol) for their very useful and motivating discussions that helped me to develop the idea behind BARM.

I am also grateful for the support that I got in the Software Campus program. The Deutsche Telekom AG (DTAG)/Telekom Innovation Laboratories (T-Labs) supported my work within the context of this program. My Software Campus project was funded by the German Federal Ministry of Education and Research (grant number 01IS12056). Project results are an important part of my thesis. Hence, I would like to thank the Software Campus team for organizational support, DTAG/T-Labs for the industrial mentoring, TU Berlin/SecT for the academic mentoring, and the Federal Ministry of Education and Research for the financial support.

There are many more people who helped me in various ways during my time as a Ph.D. student. I cannot list all of them here, but I would especially like to thank the following supporters for their assistance (incomplete list in no particular order): Yacine Gasmi, Martin Unger, Kei Ishii, and Marcel Selhorst. Additionally, the Ph.D. board, i.e., Hans-Ulrich Heiß, Jean-Pierre Seifert, and the external referees Konrad Rieck as well as Volker Roth gave me valuable feedback for the final version of my thesis for which I am also very grateful.

Furthermore, I would particularly like to thank my family, especially my parents and my sister for their encouragement and love. Finally, I am deeply grateful to my fiancee. Gesche, you always greatly helped me with whatever you could. Thank you for your brilliant support, encouragement, and love!

Abstract

Adversaries can deploy rootkit techniques on the target platform to persistently attack computer systems in a stealthy manner. Industrial and political espionage, surveillance of users as well as conducting cybercrime require stealthy attacks on computer systems. Utilizing a rootkit technique means that a part of the implemented attack code is responsible for concealing the attack. Attack code that is loaded into peripherals such as the network interface card or special micro-controllers currently are the peak of the evolution of rootkits. This work examines such stealthy peripheral-based attacks on the host computer. Peripherals have a dedicated processor and dedicated runtime memory to handle their tasks. This means that these peripherals are essentially a separate system. Attackers benefit from this kind of isolation. Peripherals generally communicate with the host via the host main memory. Attackers exploit this fact. All host runtime data are present in the main memory. This includes cryptographic keys, passwords, opened files, and other sensitive data. The attacker only needs to locate such data. Subsequently, attackers can read and modify the data unbeknownst by utilizing the direct memory access mechanism of the peripheral. This allows for circumventing security software such as state-of-the-art anti-virus software and modern hardened operating system kernels.

Detecting such attacks is the goal of this work. Stealthy malicious software (malware) that is based on an isolated micro-controller is implemented to conduct an attack analysis. The malware proof of concept is called DAGGER, which is derived from *Direct memory Access based keystroke code loGGER*. The development and analysis of this malware reveals important properties of peripheral-based malware. The results of the analysis are the basis for the development of a novel runtime detector. The detector is called BARM—*Bus Agent Runtime Monitor*. This detector reveals stealthy peripheral-based attacks on the host main memory by exploiting certain hardware properties. A permanent and resource-efficient measurement strategy ensures that the detector is also capable of detecting transient attacks. Such transient attacks are possible when the applied measurement strategy only measures at certain points in time. The attacker exploits this measurement strategy by attacking the system in between two measurements and by destroying all attack traces before the system is measured. The detector represents an alternative solution for previously proposed preventive protection approaches, i.e., input/output memory management units. Previously proposed approaches are not necessarily effective due to practical issues. This fact as well as the threat posed by peripheral-based malware demand the alternative detector solution that is presented in this work. The detector not only reveals an attack, but also halts the malicious device. BARM immediately detects and prevents attacks that are conducted by DAGGER . The performance overhead is negligible. Furthermore, BARM is able to report if the host main memory is attacked by a peripheral to an external platform.

Publications Related to this Thesis

The work presented in this thesis resulted in the following peer-reviewed publications:

- *Understanding DMA Malware*, Patrick Stewin and Iurii Bystrov, DIMVA2012 Proceedings of the 9th Conference on Detection of Intrusions and Malware & Vulnerability Assessment, Heraklion, Crete, Greece, July 26–27th, 2012 ([see 123]/Chap. 4)
- Extended Abstract – *Poster: Towards Detecting DMA Malware*, Patrick Stewin, Jean-Pierre Seifert, Collin Mulliner, CCS2011 Proceedings of the 18th ACM Conference on Computer and Communications Security, 2011 ([see 126]/Chap. 4)
- *A Primitive for Revealing Stealthy Peripheral-based Attacks on the Computing Platform's Main Memory*, Patrick Stewin, RAID2013 Proceedings of the 16th International Symposium on Research in Attacks, Intrusions and Defenses (RAID), St. Lucia, October 23−25, 2013 ([see 122]/Chap. 5)

The following peer-reviewed publications were updated for Chapter 6 to consider the DMA-based malware scenario, which is the focus of this thesis:

- *Beyond Secure Channels*, Yacine Gasmi, Ahmad-Reza Sadeghi, Patrick Stewin, Martin Unger, N. Asokan, STC2007 Proceedings of the 2007 ACM Workshop on Scalable Trusted Computing, 2007 ([see 52])
- *An Efficient Implementation of Trusted Channels based on OpenSSL*, Frederik Armknecht, Yacine Gasmi, Ahmad-Reza Sadeghi, Patrick Stewin, Martin Unger, Gianluca Ramunno, Davide Vernizzi, STC2008 Proceedings of the 3rd ACM Workshop on Scalable Trusted Computing, 2008 ([see 10])

Contents

Chapter 1
Introduction

> *Most people, I think, don't even know what a rootkit is, so why should they care about it?*
>
> Thomas Hesse,
> Former President of Sony's Global Digital Business

Many people associate the term *rootkit* to attacks on computer platforms. In fact, adversaries deploy rootkits to attack computer users. Rootkit-based attacks are used to conduct industrial espionage as well as political espionage, and cybercrime [see 16, pp. 22–25]. Adversaries conduct industrial espionage to steal intellectual property of competitors to slash the cost of technology development cycles. Political espionage differs from industrial espionage. In the case of political espionage the adversaries are interested in national secrets instead of novel technology. Cybercriminals use rootkits to steal internet banking credentials, passwords, and other sensitive data. Rootkits can also be utilized for conducting persistent surveillance of end users. Rootkits are also utilized by law enforcement, as well, to perform surveillance on suspects [see 16, p. 21]. But what exactly is a rootkit? Is it a *backdoor*? Is it a *Trojan horse*? In other words, what kind of malicious payload does a rootkit contain and how is the target computer infiltrated with the rootkit?

Several definitions for the term rootkit can be found in the literature such as Bill Blunden's *The Rootkit Arsenal: Escape And Evasion In The Dark Corners Of The System* [16]. His work also evaluates the rootkit definitions of Mark Russinovich (known from the *Windows Internals* series [106]) and Greg Hoglund (author of *Rootkits: Subverting the Windows Kernel* [60]). Finally, Bill Blunden came up with his own definition [see 16, p. 12]:

> A rootkit establishes a remote interface on a machine that allows the system to be manipulated [...] and data to be collected (e. g., surveillance) in a manner that is difficult to observe (e. g., concealment).

All these definitions imply an important property exhibited by rootkits in general, namely the capability of operating stealthily. Attackers deploy rootkits to camouflage the malicious code that attacks the target computer. This answers the question about the malicious payload of a rootkit. The payload can be anything that implements

© Springer International Publishing Switzerland 2015
P. Stewin, *Detecting Peripheral-based Attacks on the Host Memory*,
T-Labs Series in Telecommunication Services, DOI 10.1007/978-3-319-13515-1_1

malicious behavior from the user's perspective. This malicious behavior can also be a backdoor. A backdoor is used to bypass security mechanisms such as authentication requests to gain access to a computer system. A backdoor can also provide an attacker with remote access to a computer. From the attacker's point of view it makes sense to hide the backdoor. The backdoor should be used without the knowledge of the computer user. Hence, a backdoor can benefit from rootkit mechanisms. Another example for rootkit payload is a surveillance program that activates the microphone and camera of the target computer to stealthily monitor the computer user. A keystroke code logger that captures all keystrokes that are entered by the computer user is also a popular example for malicious payload.

However, the challenge for the attacker is the infiltration of the target computer platform. The attacker has to implement some kind of rootkit installer. A rootkit installer is commonly referred to as *dropper* [see 16, p. 9, 33]. Such a dropper can be based on one of the most popular infiltration mechanism, a Trojan horse or Trojan in short. The goal of a Trojan is to mislead the target computer in installing a desired program, feature or function. Instead, the user installs malicious payload such as a keystroke code logger or a backdoor. Such a payload is generally deployed in a highly privileged environment and camouflaged using rootkit techniques. Another popular infiltration approach is the exploitation of a security vulnerability. The rootkit installer could implement a so-called *exploit*. An exploit is attack code that utilizes a security vulnerability. So-called *zero-day exploits* are more threatening than non-zero-day exploits. A zero-day exploit utilizes a previously unknown security vulnerability, which can be advantageous for the attacker. It enables the attacker to conduct a stealthy infiltration of the target computer.

Another key rootkit property is that the rootkit code runs with the highest privileges as possible. The goal is to gain at least higher privileges than any potential detection mechanism. This allows the rootkit to control and modify the detection mechanism. At a certain point the detection mechanism will fail to detect the rootkit or the malicious payload that is camouflaged by the rootkit. This is the reason why attackers seek new and more powerful attack vectors. The more privileges the attacker has the more control of the target computer the attacker gains.

The goal of the attacker is to gain absolute control of the target computer. The *rootkit evolution* documents the arms race between attackers and the anti-malware community. Rootkits moved to more privileged execution environments compared to the original rootkit. In recent years [35, 36, 47, 134, 135], the rootkit evolution reached a new level. Attackers started to exploit the isolated execution environments of platform peripherals. Peripherals with a dedicated processor, dedicated memory, and a hardware feature to directly access the runtime memory of the host are able to camouflage malicious payload that attacks the target computer. Such attacks are supposed to be stealthy. No modern anti-virus like software that is available on the market considers the peripheral-based execution environments. Such software is executed on the host processor and usually only considers the harddisk and the main memory for storing malicious code.

1.1 Problem Statement

Malware is a threat for the confidentiality, integrity, and also for the availability of data. In the case of peripheral-based malware, the attacker can exploit the stealth potential of peripherals. Malware hidden in platform peripherals is not considered by anti-virus software. Depending on the peripheral, security software can not even access the inner workings of the device. For example, certain management controller have access to the whole host memory and offer remote administration features. To prevent abuse, the manufacturer applies protection mechanisms that thwart access to the inner workings of this execution environment.

The mechanism, which is exploited by peripheral-based malware to attack the host, is called direct memory access or DMA. In this work, we will introduce the term *DMA malware* for such classes of attacks. DMA malware has similar characteristics to rootkits. Current countermeasure approaches are unable to deal with the challenge of DMA malware. For example, mechanisms such as load-time integrity checks of the code intended to run on the peripheral does not prevent runtime attacks. The same is true for digitally signed firmware images. Another approach is latency-based attestation. This kind of attestation requires that a checksum be computed within a certain timeframe. Unfortunately, it also requires the modification of the peripheral's firmware and does not prevent transient attacks. Further approaches such as special monitoring and memory bus snooping are based on special hardware or hardware features. Preventing sensitive data from being present in the main memory also does not help. Such data can be dumped into the main memory via a DMA attack.[1]

A proposed countermeasure approach against DMA attacks is the utilization of a so-called *Input/Output Memory Management Unit* (I/OMMU). Such a management unit can restrict the access of peripherals to parts of the host main memory. Unfortunately, this technology has significant deficiencies. It was demonstrated that I/OMMUs can be attacked and circumvented [111, 146–148]. Hence, the I/OMMU is not necessarily trustworthy. Some operating systems such as Windows do not provide a device driver to support the I/OMMU. Additionally, not every chipset provides an I/OMMU. Furthermore, memory access policy conflicts cannot be handled by an I/OMMU. For example, Bulygin [25] demonstrated how to use a peripheral to reveal malware present in the host runtime memory. We use the same execution environment for our attack study in Chap. 4. If the I/OMMU is configured to allow the peripheral to scan the whole host runtime memory to reveal rootkits, then our attack code can also access the whole runtime memory to steal sensitive data, for example. Hence, this work does not rely on I/OMMUs as a countermeasure. Furthermore, I/OMMUs can introduce significant performance overhead [13, 150], which makes I/OMMUs undesirable in certain scenarios. Due to these considerations,

---

[1] Details can be found in Sect. 3.2 "Related Work–Countermeasure Approaches".

we believe that a runtime monitor that can detect malicious memory access with negligible performance overhead is missing. The absence of a runtime monitor is one of the major motivations for this work.

1.2 Research Question and Methodology

Our research interest is based on the stealth capabilities of modern x86 platforms. These capabilities are exploited by adversaries to hide malicious code as documented by the rootkit evolution, see also Sect. 2.1. This raises the question whether or not undetectable software can exist at all. To examine this question we consider the next logical step in the evolution of rootkits, i.e., exploiting platform peripherals to attack the host runtime memory.

We developed a malware *Proof of Concept (PoC)* that is executed on an isolated peripheral. The hardware of this peripheral provides access to the host runtime memory. We implemented an attack in the form of a keystroke code logger. This means that our malware searches for the keyboard buffer of the host operating system and monitors that buffer to capture keystroke codes. The evaluation of the keystroke logger led us to a follow-up research question, i.e., is the host system able to defend itself against peripheral-based host main memory attacks? To answer this question, we implemented a runtime monitor that is executed on the host CPU. With this monitor we want to demonstrate that additional (malicious) accesses to the host main memory that originate from platform peripherals can in fact be detected. We require that the host CPU-based monitor detects malicious accesses even if it is unable to access the isolated execution environment of the malicious peripheral.

We used our malware example to derive typical properties of this class of malware. Afterwards, we exploited these properties to detect memory accesses conducted by the malware. We identified a property that every peripheral-based malware that attacks the host memory exhibits. Because of this, we consider our malware proof of concept as typical for this malware class. We implemented the host CPU-based detector to reveal illegitimate memory accesses conducted by platform peripherals via direct memory access. The goal was to implement a runtime monitor that does not only cause minimal performance overhead for the host CPU, but also prevents transient attacks.

We also consider the network interface card in the last part of our research. The network interface card could also host malware. Especially in enterprise environments it is required that a computer platform reports its status to a central administrator platform. Such a status report can be modified by malware that is executed on the network interface card. Hence, we developed an authentic reporting channel. This channel helps to reveal attacks on such a status report.

Experimental Research Environment Our experimental environment is based on Intel x86 hardware. The isolated peripheral that we use for our peripheral-based malware is *Intel's Manageability Engine* (Intel ME [79]). The Intel ME is a special micro-controller that runs a powerful platform management firmware. An administrator can use the management firmware to remotely reinstall the operating system even if the operating system is not bootable and the platform is not reachable via the operating system's network stack. The ME also works when the platform is in standby or powered off. Due to these features the manufacturer Intel established protection mechanisms that cannot be circumvented without significant effort. The ME is isolated from the host system. The Intel ME environment is completely isolated from the host, whereas other peripherals can be accessed via debug registers and other mechanisms.

From a detector's point of view the ME is the worst case execution environment for hosting peripheral-based malware. The host CPU is unable to access the ME environment. We use this worst case environment for our research. We infiltrate the ME environment by applying an exploit that only works with a certain chipset.[2] Please note, this work does not aim to find undiscovered security vulnerabilities. We reused a known security vulnerability to set up our experimental environment due to the lack of an appropriate Intel developer board.

1.3 Impact of Thesis Contributions

To conduct industrial espionage or steal online banking credentials, for instance, attackers demand stealthily operating malware. Peripheral-based malware ensures that the attack remains to be undetectable. Peripherals that fulfill the requirements for stealthy malware operation are present in almost every modern computer platform. Peripherals such as video cards, network interface cards, and management controllers are part of desktop computers, server systems, and other computer terminals. Mobile phones and tablet computers also have peripherals with a dedicated processor, memory, and direct access to the host runtime memory. This means that all modern platforms are susceptible to peripheral-based malware attacks. Such malware is executed in an isolated execution environment and outside the scope of anti-virus software and security mechanisms set up by the operating system kernel. Due to the lack of a detector for peripheral-based malware and the lack of similar functionality in anti-virus software, the contributions of this thesis can have impact on the mentioned computer devices and their users. We summarize the main contributions of this thesis in the following:

- **DMA malware study**: We define DMA malware to be able to distinguish different DMA code. Such malware is executed on a peripheral and able to attack the host via direct memory access. We develop a proof of concept DMA malware implementation that is able to conduct a stealthy attack using an isolated peripheral. Our proof of concept is called DAGGER, which is derived from *DmA-based keystroke loGGER*. DAGGER can attack different host operating systems. DAGGER

---

[2] The exploit is only applicable to Intel's Q35 chipset with a certain BIOS version in place. Intel closed the corresponding security vulnerability by providing a BIOS update.

highlights how efficient and effective DMA malware is in practice. We identify the core properties of DMA malware to learn the properties of such malicious software. These properties are the basis for a DMA malware detector. In a first experiment we provide evidence that DMA side effects exist. We demonstrate how such an effect can be measured using common host CPU features. This is a first step for the development of a DMA malware detector. (see Chap. 4)

- **Detecting DMA malware**: We developed a monitor that detects DMA malware by comparing actual memory bus activity with expected memory bus activity. Our method is able to determine and compare actual bus activity without any firmware or hardware modification. The detector is based on a feature that implements permanent runtime monitoring and runs on the host CPU. We implemented and evaluated a PoC that we call *Bus Agent Runtime Monitor* (BARM). Our monitor implements a monitoring strategy that considers transient attacks. It does only cause negligible performance overhead. BARM can detect and halt DMA malware immediately. (see Chap. 5)

- **Authentic platform state reporting that excludes DMA malware**: We demonstrate that our detection method is also suitable in scenarios where a computer platform has to report its status to a central administrator platform. We establish an authentic reporting channel that reveals attacks conducted by malware executed on the network interface card. This means that we enhance BARM to reveal *Man-in-the-Middle* (MitM) attacks and to prevent relay attacks conducted by the network interface card. We implemented a channel to securely transmit the platform state information to an external computer. The platform state information enables a remote party to evaluate BARM measurement results. This means that the remote party can determine if its counterpart has been attacked by DMA malware. Our channel considers the host CPU as the channel endpoint and not the complete target platform. This excludes the network interface card from being part of the endpoint. We enhance BARM to account for memory bus activity that is caused by the network interface card. The enhanced BARM utilizes *OpenSSL* to implement the authentic reporting channel. We also modify the TLS handshake protocol to already account for platform state information in the very beginning of the communication session. Our modifications are still compliant to the TLS specification. (see Chap. 6)

A more detailed elaboration can be found in the corresponding chapters.

1.4 Structure of the Thesis

According to our methodology we structured this thesis as follows. In the next chapter we will introduce the required technical background, preliminaries as well as assumptions. The target platform for our evaluation is a modern Intel x86 based system, see Sects. 2.2, 2.3, 2.4, 2.5, and 2.6. These sections introduce the most important terms regarding the target platform, especially the host CPU, *Direct Memory Access*

(DMA) as well as bus master, and *Input/Output Memory Management Unit*. We also introduce our assumptions and the resulting trust and adversary model in Sect. 2.7. Chapter 3 covers related work. Since we consider both, the attack as well as attack detection and protection, we have to elaborate related work in both areas. Related works regarding DMA attacks are described in Sect. 3.1. Section 3.2 presents previous works that consider countermeasure approaches. Furthermore, we want to enable our target platform to report its status regarding DMA-based malware to an external platform. To do so, we require a communication channel that reveals MitM attacks of the network interface card. This is necessary, since we also consider network interface cards as dedicated hardware that can hide DMA attack code.

We conducted a study of DMA malware and present the results in Chap. 4. A definition for DMA malware is given in Sect. 4.1. In Sect. 4.2 we present DMA malware core functionality. The design and implementation of our DMA malware is presented in Sect. 4.3. Section 4.4 describes the evaluation of DAGGER, Sect 4.5 considers countermeasures and discusses in particular I/OMMU issues. In the same section is demonstrated how we were able to exploit these properties to demonstrate first DMA side effects. Since the host CPU is unable to directly realize illegitimate memory accesses conducted by compromised peripherals we try to provoke a side effect that occurs when a peripheral accesses the main memory.

The evidence of DMA side effects presented in Chap. 4 is the motivation for the runtime monitor that we introduce in Chap. 5. In Chap. 5 "A Primitive for Detecting DMA Malware" we demonstrate how DMA side effects can be exploited to develop a detection tool. We define a general detection model that helps us to build a detection tool, see Sect. 5.1. Afterwards we present a PoC implementation based on the popular Intel x86 platform in Sect. 5.2. We evaluate our implementation in Sect. 5.3. We also test BARM with the DMA malware that we developed in Chap. 4. Finally, BARM exploits the fact that our DMA malware has to search for valuable data that causes a certain amount of bus transactions.

In Chap. 6 we enhance our detection tool to implement an authentic state reporting application. The application sends BARM measurements to an external platform. The goal is a secure communication channel that excludes malware, which runs on the network interface card, from conducting a MitM attack. In Sect. 6.1 we present a model to negotiate an authentic reporting channel. We require a secure channel such as TLS that is bound to the actual communication endpoint, i.e., to the host CPU. Our PoC implementation of our authentic reporting application is based on OpenSSL, see Sect. 6.2. This implementation section also describes the BARM enhancements that are required to consider the network interface card. The evaluation of our implementation is presented in Sect. 6.3. We also test our network related BARM enhancements with our DMA malware DAGGER. Authentic reporting channel security considerations are discussed in Sect. 6.4. Our conclusions of this thesis as well as future work are presented in the last chapter, Chap. 7.

Chapter 2
Technical Background, Preliminaries and Assumptions

Putting a computer in front of a child and expecting it to teach him is like putting a book under his pillow, only more expensive.
Joseph Weizenbaum,
German/American Computer Scientist

Although it is beneficial, in order to understand our later material, to know many details about modern computer architecture, it would be unrealistic to explain all these subtle details here. Thus, we refer the reader to literature [see 54, 59, 118, 127] for a thorough treatment of this topic. We limit this chapter to the most important terms that are necessary to understand this work. We start with the *rootkit evolution*. This evolution highlights why the technical background that is presented in the following sections helps to understand this work.

2.1 The Rootkit Evolution

On the popular x86 platform the power of a rootkit strongly correlates to the execution environment, i.e., user-mode (ring 3) or kernel-mode (ring 0), for example. Modern x86 processors provide so-called protection rings to distinguish between different privileged execution environments, see Fig. 2.1. An analysis of the rootkit evolution reveals that attackers discovered new and more powerful execution environments on x86 platforms. The following paragraphs summarize different kinds of rootkits, i.e., user-mode, kernel-mode, virtual machine based, system management mode, firmware-based, and peripheral-based rootkits. This overview represents the rootkit evolution and demonstrates how the term rootkit changed in the recent years.

User-mode rootkits utilize simple techniques. The basic idea is to camouflage the rootkit as normal software [129]. For example, the attacker adds the desired malware functionality to a common software tool that is executed in user mode with super-user/*root* privileges. The modified tool replaces the original tool on the target platform. User-mode rootkits are considered as the starting point in the rootkit evolution. The name is derived from the privilege level that is given by the super-user

© Springer International Publishing Switzerland 2015
P. Stewin, *Detecting Peripheral-based Attacks on the Host Memory*,
T-Labs Series in Telecommunication Services, DOI 10.1007/978-3-319-13515-1_2

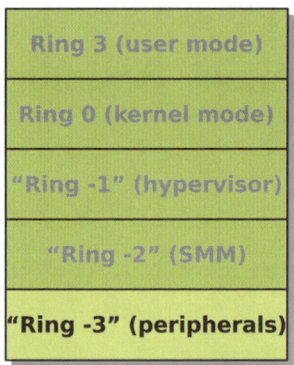

Fig. 2.1 "Ring -3" environment compared to other rootkit environments on the x86 platform. Please note, ring 3 and ring 0 are implemented in hardware (host CPU). The terms "ring -1", "ring -2", and "ring -3" are used to emphasize the power of the corresponding execution environments. They are not implemented in hardware

root. User-mode rootkits can be discovered by special detection software running in kernel-mode.

Kernel-mode rootkits are based on an advanced technique to hide the rootkit using operating system kernel components [60]. Kernel-mode rootkits modify the kernel, or to be more precise, kernel code (for example system calls) or kernel data. Kernel modifications change the kernel behavior to enforce certain stealth capabilities to hide malicious activities [see 129], e.g., a keystroke code logger. The rootkit executed in kernel-mode is immune to techniques that reveal user-mode rootkits.

Much more powerful rootkits to control a computer system are *Virtual Machine Based Rootkits* (VMBR) such as *SubVirt* [77] and *Blue Pill* [108]. A controlling instance that is called hypervisor or *Virtual Machine Monitor* (VMM) is normally used to host guest operating systems in *Virtual Machines* (VMs). A VMBR exploits the VMM environment to host the operating system of the target computer in a virtual machine. Since the operating system kernel is executed on top of the VMM environment, VMBRs can be considered to be run in "ring -1". Thus, a malicious controlling instance is placed between hardware and operating system. VMBRs are hard to install. Conversely, VMBRs are also hard to detect. Blue Pill can host the target operating system on-the-fly, i.e., without a shutdown or reboot.

Another powerful execution environment for rootkits is the *System Management Mode* (SMM). SMM is a special high privileged processor mode that executes special system software. It can also be exploited to implement so-called SMM-based rootkits. Code executed in SMM runs with the highest host CPU privileges. This means that a SMM-based rootkit runs with more privileges than the operating system kernel and a hypervisor. Hence, SMM-based rootkits can be considered to be executed in protection "ring -2" [145]. In 2008, Embleton et al. [49] and Wecherowski [144] demonstrated how SMM can be used for rootkits. SMM code is stored in firmware, i.e., SMM rootkits can be considered as a special case of firmware rootkits.

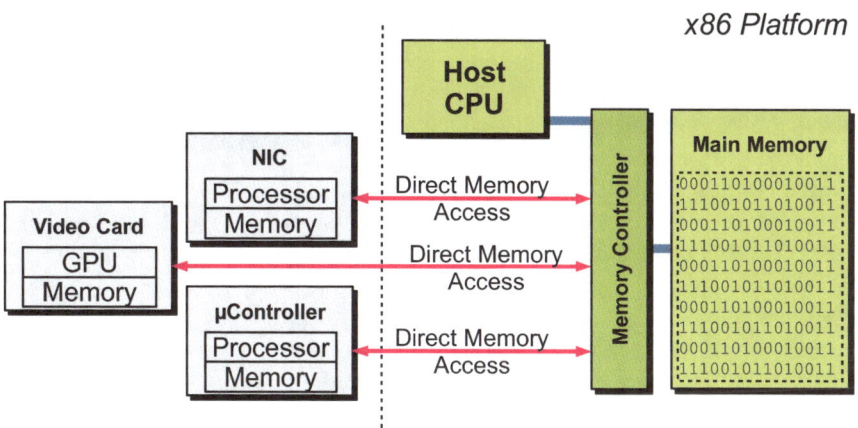

Fig. 2.2 Overview of dedicated isolated hardware potentially exploitable by rootkits. Rootkits hidden in peripherals can directly access the main memory of the computer platform. Hence, they can steal sensitive data, such as the harddisk encryption key, the video telephony session key, online banking credentials, passwords, open files, etc. It is also possible that such rootkits modify data in the main memory

Firmware-based rootkits are also quite powerful. Deploying rootkits in firmware is very difficult, but not impossible. Firmware is special low-level software that is stored on flash memory. The *Basic Input/Output System* (BIOS) is an example of firmware that is stored on flash memory on the x86 platform. A firmware-based rootkit is not deployed on a disk. Thus, it is very difficult to detect and to remove the malicious software. An attacker can use the rootkit to control the computer hardware and to attack the operating system, even if the user reinstalls the operating system. Heasman [56] demonstrated how to implement and detect a BIOS-based rootkit at Black Hat Federal 2006. Heasman [57] continued this research. Further BIOS firmware attacks that can be the basis for a rootkit were presented by Wojtczuk and Tereshkin [149], Loukas [84, 85], and Ortega and Sacco [97, 98]. Brossard [21, 22] also demonstrated that *hardware backdooring is practical*. The author exploits the open source BIOS coreboot[1] and related tools to flash the BIOS and read-only memory of peripherals to attack a computer platform.

Rootkits hidden in firmware can also be implemented using firmware of platform peripherals. Such rootkits are *peripheral-based rootkits*. A potentially exploitable peripheral is the network card [134]. Heasman [55] also discussed how to implement and detect a *Peripheral Component Interconnect* (PCI) based rootkit deployed in expansion *Read Only Memory* (ROM) that is present on the PCI device. Peripherals are well isolated from the actual host system. Hence, the execution environments of peripherals are unconsidered by anti-virus software. This makes peripherals quite attractive for attackers, see Fig. 2.2.

---

[1] See http://www.coreboot.org/Welcome_to_coreboot [accessed 25 February 2014].

A special micro-controller that executes platform management code on a separate processor offers nice stealth capabilities and can also be used by rootkits. During the Black Hat USA 2009 conference Tereshkin and Wojtczuk [131] presented the idea to use this micro-controller for rootkits. They introduced the term "ring -3" to emphasize the stealth capabilities. Such peripheral-based rootkits are considered to be even more stealthily than SMM-based rootkits. Bulygin [25] demonstrated how to use this special micro-controller based environment to detect SMM-based and VMM-based rootkits. Since peripherals such as network interface cards communicate with the host operating system via the main memory, peripheral-based rootkits can attack the host by illegitimately reading from or writing to the host memory. The mechanism that enables memory access for peripherals is called *Direct Memory Access* (DMA, see Sect. 2.4). Due to this mechanism peripheral-based rootkits are supposed to be absolutely stealthy and undetectable. Such rootkit techniques are the focus of this work. Peripheral-based rootkits can access the host memory to steal passwords, online banking credentials, open documents, etc. that are present in the host's runtime memory via DMA. They can also infiltrate the host with further attack code such as a kernel-based backdoor [47].

Note, in this work we avoid the term "ring -3". No "ring -3" is implemented in hardware. Terms such as "ring -1", "ring -2", and "ring -3" are only used to illustrate the privilege level of the corresponding environment on the x86 platform. The lower the ring the more powerful is the rootkit. In this thesis, we will use the term malware (malicious software) because the attacks that we analyze are not executed on the host CPU. Hence, root privileges are irrelevant. The malware that we focus on has only the goal to operate stealthily in common with original user space rootkits.

2.2 Typical x86-Based System Architecture

The main components of a typical x86 system architecture as depicted in Fig. 2.3. The linkage of *Central Processing Unit* (CPU), *Memory Controller Hub* (MCH), and *Input/output Controller Hub* (ICH) is called the chipset [54]. This chipset solution is also referred to as *3-chip solution*. System memory (*Random Access Memory* or in short RAM) as well as a display adapter are connected to the MCH. The MCH controls access to memory. It can block requests to memory addresses or redirect the request to the ICH, if the destination address belongs to the ICH. Peripheral devices, such as flash memory, *Network Interface Card* (NIC), etc., are integrated into the system using the *Peripheral Component Interconnect express* (PCIe [24]) standard. This standard implements a serial interconnect for peripherals and the chipset. NICs and other add-on cards can be connected to the ICH via PCIe. Flash memory, which stores firmware such as the *Basic Input/Output System* (BIOS [see 54, p. 369]), is also connected to the ICH.

Please note, Intel introduced a so-called *2-chip solution* with the *Intel 5 Series chipset* [121, p. 15]. 2-chip solution means that the MCH functionality moved into the host CPU and is called *Integrated Memory Controller* (IMC [32, p. 14]). The IMC

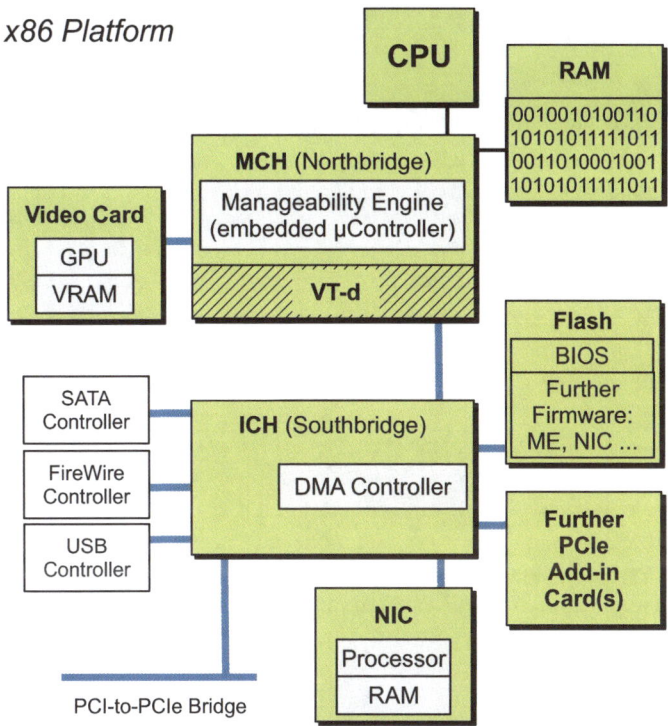

Fig. 2.3 x86 chipset and peripheral components. The chipset components are the *Central Process-ing Unit* (CPU or host processor), the *Memory Controller Hub* (MCH, also known as northbridge) and the *Input/output Controller Hub* (ICH, also known as southbridge). Peripherals do not belong to the main chipset

is the controlling instance that controls memory accesses just as the former MCH. The ICH was renamed to *Platform Controller Hub* (PCH [68]). The experiments conducted in this thesis are based on the 3-chip solution.

Further controller devices connect other formats, such as *Universal Serial Bus* (USB [8]), *FireWire* (FW [6]), or *Serial Advanced Technology Attachment* (SATA [7]), via PCIe to the system. Legacy PCI devices are connected to the PCIe architec-ture via a so-called *PCI-to-PCIe bridge* [24]. In laptop computers *Personal Computer Memory Card International Association* (PCMCIA)/*ExpressCard* [139] devices are integrated into the system utilizing PCIe. The host CPU is not necessarily the only processor in the system. The video card, for example, supports a *Graphics Processing Unit* (GPU) to efficiently modify computer graphics. Data to be processed is stored in *Video RAM* (VRAM), that is separated from normal system RAM. Other devices with similar properties are NICs and *Intel's Manageability Engine* (ME [79]) in the platform's MCH. They also utilize separate processors as well as separate RAM to execute firmware.

2.3 Intel x86 Based Host Central Processing Unit

The Intel x86 *Central Processing Unit* (CPU) was announced in 1978 [see 59, Appendix K.3]. Since then, the x86 CPU has been enhanced and nowadays x86 processors consist of several units to support proper features for different computing tasks. Modern extensions are floating point unit, *Single Instruction operating on Multiple Data items* (SIMD [117, p. 524]), *Streaming SIMD Extensions* (SSE [117, p. 748]), x64 [58, p. 351], *Physical Address Extension* (PAE [69, pp. 2–23]), multi-level caches (L1, L2, L3 cache [59, p. 117]), *Performance Monitoring Units* (PMU [see 104, p. 429]) and hardware support for virtualization as described by Grawrock [54]. A modern x86 CPU usually consists of multiple cores [see 59, p. 117]. These cores provide registers of different bit sizes, i.e., from 16 bit up to 512 bit [see 70, Sect. 1.2.1].

To offer protection mechanisms the CPU supports a privilege model via the so-called protection mode. The model provides different privilege levels also known as rings to separate certain software running on top of the hardware. Four rings are available if the processor is in protected mode. Ring 0 is the most privileged ring ring 3 has the fewest privileges. The operating system is executed in ring 0. Thus, it is separated from application software running in ring 3. Ring 1 was considered for device drivers and ring 2 for services, though in practice ring 1 and 2 are not used [54, p. 41].

System Management Mode (SMM [69]) is another processor mode only available for system firmware. That mode was introduced in x86 architectures to implement higher energy-efficiency by, e.g., powering down unused disks and to control system hardware by, e.g., turning on fans and shutting down systems when temperature limits are reached. SMM is triggered by an interrupt, i.e., the *System Management Interrupt* (SMI). SMI handler code is loaded from flash memory by the BIOS into the *System Management RAM* (SMRAM) early in the system initialization. To prevent modifications of the SMI handler code from other processor modes than SMM, the chipset provides a special bit that is called D_LCK. The D_LCK bit is set to protect the SMI code after loading it into SMRAM. If the D_LCK bit is set no alteration of SMRAM content is possible.

When an SMI triggers SMM, the current executed program is interrupted and the processor state will be saved. Afterwards, the processor executes the SMI handler code. When the execution of the handler code has been completed, the saved processor state is restored. After the processor switches back from SMM to the previous processor mode the interrupted program can continue to operate. Note that the previous processor mode has lost CPU cycles/time, since both processor modes cannot be executed simultaneously. SMM can be considered to be a separate execution environment. SMRAM is a separate address space and only accessible when the processor is in SMM. In other words, the OS has no access to SMRAM. Furthermore, privileges in SMM are not restricted, code executed in SMM can call all I/O as well as system instructions.

Hardware virtualization extensions in x86 are called *Intel Virtualization Technology* (Intel VT) on Intel platforms [54]. Virtualization mechanisms are used to run multiple OSes or applications isolated from each other on a single hardware platform in parallel. A controlling instance called a *hypervisor* or a *Virtual Machine Monitor* (VMM) hosts guest OSes in *Virtual Machines* (VMs). Modern x86 CPUs provide a special instruction set called VT-x. VT-x is part of Intel VT and is intended to support hardware virtualization. This hardware support offers two special CPU operations: VMX root operation and VMX non-root operation. A VMM is run in VMX root operation. VMs running on top of the VMM are executed in VMX non-root operation controlled by the VMM. Both operation modes support their own protection rings, four rings each. Thus, software of the guest system (kernel, drivers, applications, etc.) can be run in the designated privilege level. The protection rings in VMX non-root operation are considered to be unprivileged, since these rings are controlled by the VMM running in VMX root operation. The four rings of the VMX root operation mode are privileged. Usually, the VMM uses only the most privileged ring. This ring is often called "ring -1" to emphasize that it controls the unprivileged rings 0–3.

The x86 micro-architecture also implements a pipelining concept with special execution optimization features, such as branch prediction and out-of-order execution [118, p. 329ff] [127, p. 93ff]. The execution pipeline works with micro-operations, i.e., computations that are implemented as stylized atomic units. Intel architecture instructions are translated into micro-operations [118, p. 331]. For out-of-order execution a so-called *Reorder Buffer* (ROB [118, p. 333]) is required to keep track of renamed registers. Register renaming occurs during out-of-order execution. Registers used in micro-operations are renamed by utilizing the *Register Alias Table* (RAT [118, p. 333]) that is also referred to as the *Register Allocation Table* (RAT [see 127, p. 100]).

PMUs are implemented in the form of *Model-Specific Registers* (MSR [69, Sect. 9.4]) that enable software developers to count micro-architecture related events. This helps programmers to write optimal code for a certain CPU micro-architecture [104]. For example, the MSRs can be configured to count cache misses, RAT stalls, and branch mispredictions that occur when executing code [69, Chaps. 18/19]. The PMU registers that count events are also referred to *Performance Counter* or *Hardware Performance Counter* (HPC). They are only available in ring 0. Another special purpose register that is related to performance measurements is the so-called *Time Stamp Counter* (TSC [69, Sect. 17.12]) register. The TSC register can be used to count CPU cycles after a platform reset. Access to the time stamp counter register as well as to the performance monitoring unit registers from different privilege levels can be controlled via the x86 control register 4 (CR4) [see 69, Chap. 2].

A special input/output (I/O) feature to exchange data with peripherals is the concept of I/O-mapped I/O via ports (I/O ports [117, p. 70, 341]) that is provided by the x86 CPU. This concept is complementary to memory mapped I/O (also supported by x86 systems [117, p. 343]) where memory as well as registers of peripherals are mapped into the memory address space of the host CPU. Peripherals also communicate with the host CPU via interrupts to signal that new data is available, for example [117, p. 252]. To communicate with the host system, peripherals can also use the concept of direct memory access. In this case the peripheral does not communicate directly with the host CPU, see Sect. 2.4.

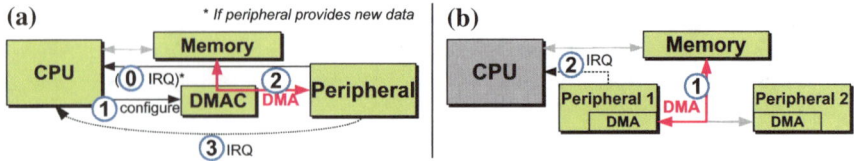

Fig. 2.4 Third-party and first-party DMA. **a** Third-party DMA: The host CPU is required to (*1*) configure (source and destination address) the central DMA controller via I/O ports to (*2*) perform a DMA transfer. The host CPU is (*3*) interrupted when the DMA transfer has been finished [31, p. 454]. Hence, the host CPU is aware of a third-party DMA transfer.—**b** First-party DMA: The peripheral device can (*1*) configure its own DMA engine. The device acts as bus master (see Sect. 2.5) to get control of the system bus to perform a DMA transfer. The device *can* interrupt the host CPU when the device (*2*) has completed the transfer. The transfer also works if the device does not interrupt the host CPU at the end of the DMA transfer. In this case the CPU is unaware of the DMA transfer

2.4 Direct Memory Access

PCIe supports *Direct Memory Access* (DMA) for peripherals, or to be more precise for dedicated hardware such as video cards, NICs, and management controller. DMA enables fast memory access without the involvement of the host CPU. The aim of DMA is to remove the burden from the host CPU. DMA allows peripherals to gain access to the whole host memory bypassing the CPU. The CPU can perform other tasks while DMA transfers occur. Peripherals can have their own engines to perform DMA. This kind of DMA is called first-party DMA [133, p. 428]. Another mechanism is third-party DMA [133, p. 428] where a central *DMA Controller* (DMAC, see Fig. 2.3) is necessary to provide legacy devices (e.g., devices based on the *Industry Standard Architecture* (ISA [116]) format) without DMA engines with fast memory access. It is also integrated in modern platforms [64, p. 128].

Figure 2.4 highlights an important difference regarding stealthy operation between third-party and first-party DMA. When using third-party DMA the host CPU is aware of the DMA transfer, because the peripheral needs the host CPU to configure [see 31, p. 454] the DMAC via I/O ports[2] (see Sect. 2.3). When using first-party DMA the host CPU is not *necessarily* aware of the transfer. Note, a DMAC or a DMA engine can only access host memory addresses, but not host CPU cache, host CPU registers, or the harddisk, for example. The latter implies that data swapped out from runtime memory to the harddisk is not accessible by a DMA engine, either.

---

[2] See the Linux source code files `arch/x86/include/asm/dma.h` and `arch/x86/include/asm/io.h`, for example.

2.5 Bus Master

A computer platform has several bus systems, such as PCIe and *Front-Side Bus* (FSB). Hence, a platform has different kinds of bus masters depending of the bus systems, see Fig. 2.5. A bus master is a device that is able to initiate data transfers (e.g., from an I/O device to the main memory) via a bus [58, Sect. 7.3]. A device (CPU, I/O controller, etc.) that is connected to a bus is not per se a bus master. The device is merely a *bus agent* [1, p. 13]. If the bus must be arbitrated a bus master can send a bus ownership request to the arbiter [9, Chap. 5]. When the arbiter grants bus ownership to the bus master, this master can initiate bus transactions as long as the bus ownership is granted. Note, this procedure is not relevant for PCIe devices due to its point-to-point property. PCIe requests are not required to be arbitrated and therefore, bus ownership is not required. The bus is not shared as it was formerly the case with the PCIe predecessor PCI.

Nonetheless, the bus master capability of PCIe devices is controlled by a certain bit, that is called *Bus Master Enable* (BME). The BME bit is part of a standard configuration register of the peripheral and is usually set by the corresponding device driver that is executed on the host CPU. The MCH (out of scope of PCIe) still arbitrates requests from several bus interfaces to the main memory [63, p. 27], see Fig. 2.5. The host CPU is also a bus master. It uses the *Front-Side Bus* (FSB) to

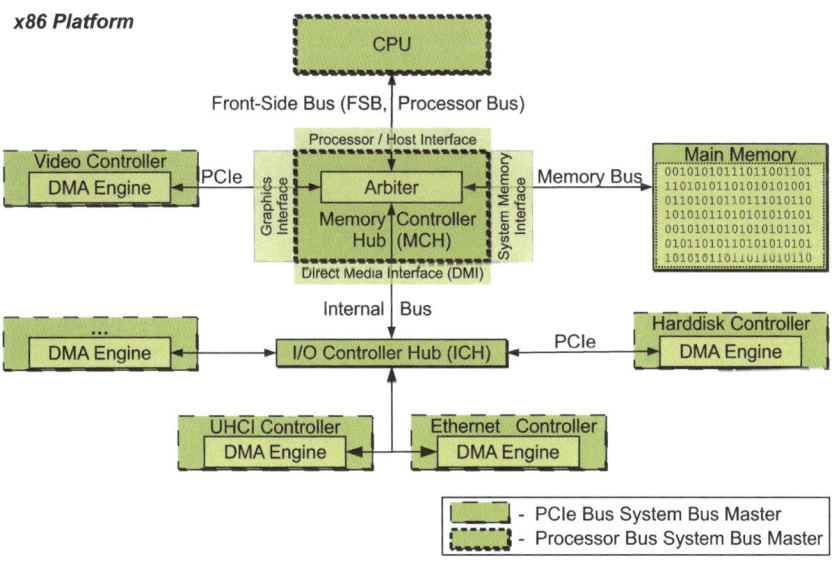

Fig. 2.5 Bus master topology. Bus masters access the memory via different bus systems (e.g., PCIe, FSB). The MCH arbitrates main memory access requests of different bus masters. (based on [23, p. 504][24][58, Section 7.3][63, Section 1.3][64])

fetch data and instructions from the main memory. I/O controller (e.g., ethernet, harddisk controller, etc.) provide separate DMA engines for I/O devices (e.g., USB keyboard/mouse, harddisk, NIC, etc.). This means that when the main memory access request of a peripheral is handled by the MCH, PCIe is not involved at all.

2.6 Input/Output Memory Management Units

Intel introduced a technology called *Intel Virtualization Technology for Directed I/O* (VT-d, [2]) as one of several building blocks to provide hardware supported virtualization for x86 systems. VT-d can be considered as an *Input/Output Memory Management Unit* (I/OMMU) to efficiently assist virtualization requirements, such as reliable isolation of virtual machines running on a virtual machine monitor. VT-d is mainly used in conjunction with virtualization solutions. With VT-d, system software, that means a hypervisor or an OS, can create memory protection domains. For example, isolated subsets of physical memory can be assigned to a virtual machine or to memory of an I/O device driver. An I/O device that is not assigned to a protection domain has no access to physical memory of that domain. These access restrictions are realized using address translation tables. System software configures so-called *DMA Remapping* (DMAR) engines provided by Intel VT-d. Such an engine maps a memory request, for example triggered by an I/O device, to physical memory. VT-d can block a memory request, if the device is not assigned to the protection domain. Please note, an activated I/OMMU can introduce significant performance overhead for the host CPU [13] [88, p. 29] [150]) with the result that the utilization of this technology is often avoided.

To enable system software to configure DMAR engines, the BIOS is required to load corresponding information in the form of *Advanced Configuration Power Interface* (ACPI [44]) tables into the main memory. System software can use this information (e.g., number of DMAR engines) to set up protection domains. Please note, storing the ACPI tables in the main memory raises a serious security threat. These tables are accessible via direct memory access and can be modified as described by Wojtczuk et al. [148] and Sang et al. [111]. System software that is responsible to configure the DMAR engines correctly might fail if this vulnerability is exploited by an attacker.

2.7 Trust and Adversary/Attacker Model

The attacker model provides a description for a stealthy DMA attack scenario. The attacker is able to infiltrate dedicated hardware present in a computer platform with malicious payload *remotely*. This can be carried out via an OS or firmware related zero-day exploit [see 47, for example]. We assume the attacker is able to attack the target platform during runtime. This can not only be done remotely using a firmware exploit, but also via a remote firmware update mechanism as demonstrated by Duflot [45] and by Triulzi [135], respectively. Alternatively to the described remote exploitation, the attacker can also infiltrate the peripheral before the supposed owner gains and deploys the peripheral on the target platform.

The dedicated hardware supports first-party DMA as described in Sect. 2.4 and accesses the main memory via the memory bus, see Fig. 2.5. We assume that the target computer platform has usual up to date defense mechanisms such as anti-virus software and a host firewall. The platform user does not apply additional hardware such as a hardware firewall to protect the computer platform. We assume that only a stealthy attack can result in a successful attack. Hence, the attacker wants to hide the attack by using the stealth potential of dedicated hardware. Attacks on the main memory (i.e., confidentiality and integrity violations) only originate from peripherals via DMA. The attacker does not implement an attack that requires a cooperation between peripheral and host to increase the probability of a stealthy attack. We further assume that the attacker ensures that an integrity violation (memory write access) does not result in an attack revelation. Additional hardware would decrease the probability of a successful stealthy attack significantly. Most likely, the attacker aims on stealing data, e.g., to conduct industrial espionage or to acquire online banking credentials, etc. To do so, the attacker wants to read data from (confidentiality violation) or write data to (integrity violation) the main memory via DMA.

We consider a computer platform as trustworthy if it conforms to the applied security policy, that means in our case no DMA-based malware is attacking the host platform by reading from or writing to the platform's main memory via DMA. We rely on a minimal *Trusted Computing Base* (TCB [37, p. 66] [99, p. 8]) that consists of the host CPU and the RAM chip hardware as well as the communication path in between (front-side bus, memory controller hub, memory bus). Software (system software as well as application software) executed on the host CPU, is in a trusted state before the platform is under attack. This means that software is loaded as well as started correctly and behaves as expected. We do not count on preventive approaches such as I/OMMUs due to the security issues mentioned in Sect. 2.6.

Chapter 3
Related Work

> *The hacker mindset doesn't actually see what happens on the other side, to the victim.*
>
> Kevin David Mitnick,
> Security Professional

Since we determined in our methodology to consider both, the attack as well as attack detection and mitigation, we have to elaborate on related work in both areas. Furthermore, we want to enable our target platform to report its status regarding DMA-based malware to an external platform. To do so, we require a communication channel that reveals *Man-in-the-Middle* (MitM) attacks of the network interface card. This is necessary, since we also consider the NIC as dedicated hardware that can hide the attack code.

3.1 DMA Attacks

Direct memory access can be a sufficient approach to conduct stealthy attacks on the host system. Our work analyzes attacks that implement malware functionality and apply rootkit/stealth capabilities during runtime. In the worst case the DMA-based malware will also survive platform reboots and standby as well as power off modes. In the following we distinguish between peripherals that can be connected to the host platform from the chassis outside and peripherals that are directly connected to the chipset.

3.1.1 Devices Connectable from the Outside

Since 2004 several DMA attacks using additional hardware such as USB devices [90], special PCMCIA cards [11, 61], and FireWire devices [17, 42, 43] were presented. The attack demonstrated by Maynor [90, p. 55ff.] exploits a Motorola mobile phone

© Springer International Publishing Switzerland 2015
P. Stewin, *Detecting Peripheral-based Attacks on the Host Memory*,
T-Labs Series in Telecommunication Services, DOI 10.1007/978-3-319-13515-1_3

to infiltrate the target machine with attack code via USB. The attack reveals itself by displaying a window on the screen of the target platform. Hence, the attack is a proof of concept rather than fully operative malware.

Dornseif [43] and Dornseif et al. [42] demonstrated how to exploit an Apple iPod that is connected via FireWire to the target to conduct a DMA attack. The authors mention that they can copy the screen content, strings, and key material using DMA reads. Furthermore, with DMA writes, the authors can change the screen content, conduct a privilege escalation attack, and inject code into the runtime memory of the host. Boileau [17] also covered a FireWire-based DMA attack. The author was able to attack a Windows XP based laptop computer. In 2007, Piegdon and Pimenidis [101] published another FireWire related DMA attack paper. They described how to steal private SSH keys as well as to inject arbitrary code. The injected code implements interactive access to the target machine with administrator privileges. The authors had to search data structures that are used by the host CPU to implement virtual address space for processes running on the host CPU. Blass and Robertson [15] described *Tresor-Hunt*, another FireWire-based attack to trick harddisk encryption mechanisms. To be more precise, host CPU bound encryption mechanisms are attacked. CPU bound means that key data is never released to the main memory. That data is kept in host CPU registers. The basic idea of Tresor-Hunt is to inject code into kernel space (an interrupt handler is hooked). That attack code dumps the key data from the processor registers into the main memory where it can be captured via DMA. Blass and Robertson [15] use FireWire to dump the physical host memory. Then, they scan the whole dumped memory for the interrupt descriptor table to hook an interrupt handler that will eventually release the encryption key.

Hulton [61] presented how to use a *Field-Programmable Gate Array* (FPGA) peripheral that is connected via cardbus to the target platform to capture passwords and secret keys present in the main memory. Furthermore, Hulton's FPGA device is able to unlock screensaver screen locks. Aumaitre and Devine [11] also described an attack that is based on an FPGA on a PCMCIA card. It can also be used to unlock screensavers and to execute arbitrary code. To find the target memory address in the host memory the authors apply a signature scan in all physical memory pages.

The project documented by Breuk and Spruyt [18, 19] aims to integrate DMA attacks into exploitation frameworks. The authors discuss PCI, FireWire, USB, SATA, DisplayPort, Thunderbolt, and PC Card (i. e., PCMCIA, Cardbus, Express-Card). Their proof of concept is based on FireWire. To find the target address in the host's runtime memory, the authors implemented a signature scan that is applied to all memory pages. The *Inception* tool is able to attack the target platform via "FireWire, Thunderbolt, ExpressCard, PC Card and any other PCI/PCIe interfaces" [87]. The tool can, amongst other things, dump the main memory, unlock the system, and conduct a privilege escalation attack on Windows, Mac OS, and Linux based targets. Ongoing research that describes strategies to further exploit Thunderbolt for DMA attacks is presented by Sevinsky [114]. The author does not describe a concrete DMA attack via Thunderbolt.

3.1.2 Devices Firmly Established Inside the Platform Chassis

In this work we have a clear focus on stealthiness. The attacker must not need physical access to the target machine to increase the probability of stealthy infiltration. Hence, the attack devices presented in Sect. 3.1.1 are not considered by our trust and adversary model, see Sect. 2.7. We focus on attacks that originate from platform peripherals. This section considers DMA attacks that originate from platform peripherals such as special management controller, network interface cards, and video cards.

Tereshkin and Wojtczuk [131] demonstrated that the DMA engine of Intel's ME can be used to write to host memory. The authors described a vulnerability that allows to inject code into the ME environment. The code of Tereshkin and Wojtczuk did not implement any malware behavior. It reveals itself by writing to a known hard coded host memory address. Hence, this approach implements a proof of concept and no real malware functionality. We use Intel's ME for our attack study, see Chap. 4. Our DMA-based attack implements fully operative malware in the form of a keystroke code logger that is executed in the manageability engine environment.

The network interface card based attacks described by Duflot et al. [47], Delugré [35, 36] focus on stealthy attacks, malware functionality, and rootkit capabilities. The attack presented by Duflot et al. [47] exploits a vulnerability in the firmware of a NIC during runtime. The compromised NIC is used to attack the host system by adding a backdoor. The authors described how the host could access the NIC internal memory. This offers a possibility to detect the attack code using code executed on the host CPU. As far as we know no anti-virus like software makes use of this. It should be mentioned that the host access to the NIC internal memory is not a common feature. For example, the runtime memory of the Intel ME environment that we use for our attack study (see Chap. 4) is not accessible by the host. The work published by Delugré [35, 36] is quite similar to the work published by Duflot et al. [47]. Both attacks use the same NIC model. The malware presented by Delugré [35, 36] aims to implement rootkit capabilities.

Triulzi [134, 135] presented a stealthy secure shell that offers memory inspection using DMA. A combination of NIC and video card is used to hide the shell. The shell is installed by reflashing firmware remotely. NIC and video card communicate via PCI-to-PCI transfers. The author proposed to count PCI-to-PCI transfers as a countermeasure, but it was not demonstrated how this can be implemented. Other video card related work was published by Vasiliadis et al. [140]. The authors described a method to shift performance overhead away from the host CPU to the GPU of the video card. Parts of the code are still required to run on the host CPU. CPU and GPU communicate via shared memory. The performance overhead arises when techniques such as unpacking or runtime-polymorphism are used. Hence, Vasiliadis et al. [140] described GPU assisted unpacking as well as runtime-polymorphism, but no specific malware that uses DMA to attack the host system. Ladakis et al. [80] implemented a keystroke code logger that runs on a GPU. The keystroke logger is reminiscent of the keystroke code logger that we published [123].[1] They reused the same signature scan to find the keyboard buffer. Furthermore, the approach requires executing the signature scan on the host CPU in kernel mode. This Achilles' heel can be exploited to detect the attack. The authors actually require a kernel-based zero-day exploit to increase the probability of a stealthy attack. According to the authors, special debugging tools can be used to analyze processes executed on the GPU. These tools can be used to develop a countermeasure for GPU-based malware. It is also unclear what happens with the captured keystroke codes in the video card environment. Ladakis et al. [80] do not consider exfiltration.

Recently, Domburg [41] demonstrated how to install attack code on a harddisk controller. The attack code is stored on the harddisk controller flash memory and loaded into the harddisk controllor's DRAM to be executed on the processor of the harddisk controller. The author could not demonstrate how to exploit the harddisk controllor's DMA engine to attack the host runtime memory. Similar work that is also based on a harddisk controller was presented by Zaddach et al. [152]. The authors demonstrated a stealthy hard-drive backdoor. However, they attack data stored on the harddisk, i.e., they did not demonstrate how to exploit the controller's DMA engine to attack the main memory of the host system. Hence, their attack is out of scope of this thesis. We focus on stealthy attacks on the platform's main memory.

3.2 Countermeasure Approaches

Different approaches have been proposed that could be considered as countermeasures against DMA attacks. For example, measured firmware[2] is an approach to check the integrity of the firmware binary. It is assumed that the firmware does not conduct a DMA attack if the binary is unmodified. The signed firmware approach also aims at convincing the user that the firmware does not conduct a DMA attack. The idea is that firmware that is digitally signed by the vendor is trustworthy. Besides these two approaches the following sections also describe related work such as latency-based attestation, runtime monitoring, bus snooping, sensitive data protection, and the I/OMMU.

3.2.1 Measured Firmware

The *Trusted Computing Group* (TCG) [136] proposed to *attest* the peripheral's firmware at load time. To be more precise, the approach is based on an additional

---

[1] The keystroke code logger that we published [123] is the basis for our attack study in Chap. 4.

[2] In this case measurement means deriving a hash value.

chip[3] that is called *Trusted Platform Module* (TPM [99]). The TPM is similar to a smart card chip that is firmly fixed to the chipset of a computer platform. However, the TPM can store integrity measurements in the form of hash values of binary code before that code gets executed. This means that the measurement is at load time. Such measurements can be used to check if the platform is trustworthy. Current versions of Intel's Manageability Engine execution environment also utilize a so-called measured launch that enables the attestation of the peripheral's firmware using a hash value [79, Chap. 15]. Unfortunately, measurements conducted at load time do not exclude runtime attacks. Repeating the measurements during runtime causes significant performance degradation. It can also not prevent *transient attacks* where an attacker exploits the time frame between two measurements. Furthermore, it is not ensured that the host CPU is able to access all peripheral ROM components that stores the firmware code.

3.2.2 Signed Firmware

Signed firmware images do also not prevent runtime attacks. Firmware updates can only be flashed to the corresponding ROM chip if the firmware image has a valid digital signature. For example, only a BIOS firmware image that was signed by the motherboard vendor can be flashed into the corresponding ROM chip [see 79, Chap. 14]. This does not exclude runtime attacks. Attacks were demonstrated by Wojtczuk and Tereshkin [149] as well as Butterworth et al. [26].

3.2.3 Software/Latency-Based Attestation

Other attestation approaches were presented by Li et al. [82, 83], for example. These approaches are based on *latency-based attestation*, i. e., a peripheral needs not only to compute a correct checksum value. It also has to compute the value in a limited amount of time. A compromised peripheral is revealed if either the checksum value is wrong or if the checksum computation took to much time. Latency-based attestation approaches require to modify the peripheral's firmware and the host needs to know the exact hardware configuration of the peripheral to be able to attest it. Li et al. [83] also state that their approach does not work correctly when peripherals cause heavy bus traffic. They considered only one peripheral in their evaluation. Furthermore, Nguyen [96] revealed serious issues in attestation approaches as presented by Li et al. [83]. It is also unclear to which extent latency-based attestation can prevent transient attacks.

---

[3] The TCG specification does not forbid to implement the TPM in the form of firmware. Intel [see 79, p. 108] has a TPM solution based on firmware.

3.2.4 Monitoring Approaches

Another interesting approach was presented by Duflot et al. [46]. NIC adapter-specific debug features are used to monitor the firmware execution. Such features are not available for other peripherals. Another deficiency is the significant performance issue for the host (100% utilization of one CPU core). Our goal is also the development of a runtime monitor. In contrast to the monitor described by Duflot et al. [46] our monitor is required (i) to be independent of the inner workings of the peripheral and (ii) to cause significant less performance overhead, see Chap. 5.

Another runtime monitoring approach was presented by Zhang [153]. That approach is based on SMM. The author proposes to periodically check the peripherals firmware as well as configurations. Unfortunately, the author does not describe how the SMRAM that contains the monitor is protected against DMA attacks before the I/OMMU is configured correctly. The author also does not explain the checking interval. Hence, one has to assume that transient attacks are unconsidered. It is also unclear how much time is required to check all peripherals. An implementation description as well as an evaluation are missing. Thus, it is not proven that the proposed approach is applicable in practice.

3.2.5 Bus Snooping Approaches

Moon et al. [92] and Lee et al. [81] follow a different hardware-based approach. The authors propose a system that is able to snoop the memory bus to detect kernel integrity violations. The approach is able to prevent transient attacks. Unfortunately, the authors do not aim at detecting DMA attacks. Furthermore, their snoop monitor component is based on special hardware (Leon3 processor) that has the same computing power as the monitored host system (also Leon3 processor). It would be interesting to see if such a memory snooping approach can be exploited to detect DMA-based malware.

A related approach that was presented by Eckert et al. [48] considers a kind of DMA attack. The proposed system can be used to detect malware that is transfered to the host memory via DMA. Hence, the authors only consider write access from the peripheral to the host memory. The system is unable to prevent DMA read based attacks where an attacker captures cryptographic keys or online banking credentials that are present in the main memory, for example. In the described attack scenario, the authors assume that the attack code is executed on the host processor. Hence, they scan the data that is written via DMA to the host memory for malware signatures by also snooping the bus. The authors admit that their signature-based detection approach has deficiencies. The proposed system requires FPGA-based hardware and it is also unclear if they focus on first-party or third-party DMA with their implementation.

3.2.6 Sensitive Data Protection

To protect sensitive data such as cryptographic keys from memory attacks several approaches were presented. It is proposed to store sensitive data only in processor registers or in the cache, but not in the main memory [93, 94, 119, 141]. Unfortunately, Blass and Robertson [15] demonstrated how to use a DMA-based attack to enforce the host to leak the sensitive data into the main memory, see Sect. 3.1.1.

3.2.7 Input/Output Memory Management Unit

Sensitive data, which is stored in the main memory could also be protected by an I/OMMU as proposed by Duflot et al. [47] and Müller et al. [95]. As already considered in our trust and adversary model we do not rely on I/OMMUs (see Sect. 2.7). This is because an I/OMMU must be configured faultlessly [83, p. 2] and because I/OMMUs can be successfully attacked [111, 146–148]. Furthermore, I/OMMUs are not applicable due to memory access policy conflicts [123] and they are not supported by every chipset and OS. Sang et al. [112] also confirm that I/OMMUs have deficiencies. Another important point that should be considered when considering an I/OMMU as a countermeasure is that an activated I/OMMU can according to Ben-Yehuda et al. [13] and Yassour et al. [150] cause significant performance overhead.

3.3 Secure Communication Channels Considering Platform State Reporting

None of the related works presented in this section considers NICs as host for malware that can conduct a *Man-in-the-Middle* (MitM) attack. We adapt the concept of a *Trusted Channel* for this purpose [10, 52]. A trusted channel has all properties of a secure channel. Additionally, the trusted channel concept enables binding configuration data of the communication endpoint to the secure channel to ensure the authenticity (i.e., the identity as well as integrity) of the endpoint. Nonetheless, other approaches related to trusted channels exist and are discussed in the following. To prevent *relay attacks* (the attacker relays trustworthy configuration data of a third platform), it is required to implement a secure binding between the secure channel and the configuration data to be reported to the peer. Not all of the presented related works implement such a secure binding.

3.3.1 Trusted Platform Module Based Approaches

Many approaches based on *Trusted Computing* (TC [99]) as proposed by the *Trusting Computing Group* (TCG)[4] exist. Many approaches enhance existing secure channel protocols such as *Transport Layer Security* (TLS [38]) or *Internet Protocol Security* (IPsec [76]) to integrate or bind endpoint configuration data to the secure channel [120]. We also prefer to benefit from an existing secure channel protocol.

Smith [120] described how to combine platform authentication as well as user authentication to authenticate an endpoint. To do so, they introduce TLS extensions for a two-phase handshake. Unfortunately, Smith's description is not very detailed. Relay attacks as well as endpoint configuration changes are outside the scope. Sadeghi et al. [110] also introduced a trusted channel concept. Their concept is based on key transport. We prefer to use contributory key agreement. We consider key material contribution of the involved endpoints. Furthermore, the reference implementation described by Sadeghi et al. [110] uses TLS to tunnel their channel. Configuration data is not bound to the secure channel based on TLS.

The TCG developed the *Trusted Network Connect* (TNC) architecture [138]. TNC mainly addresses network access. Network authentication and policy enforcement is the focus of TNC. This is not our focus. Integrity-based configuration information are used to decide if a platform is allowed to enter the network or not. The TCG worked on a specification that extends the TLS protocol for attestation purposes (*TLS Extensions for Attestation* or TLS-Attestation in short [130, p. 51]). The document is not publicly available via the TCG website.[5] However, the TCG [137] published a document called "Binding to TLS". This document considers MitM attacks when a client requests access to a network. Another approach based on TNC is discussed by Rehbock [103]. The author extends the TNC architecture to web-based environments.

The aim of Marchesini et al. [89] is to attest the trustworthiness of web applications. They introduced an architecture that is based on the proposed *Bear platform*. That platform implements a trust model that aims to map a long-lived cryptographic key pair (certified by a certification authority) to short-lived platform configuration parts. The authors admit that their platform has some issues such as *time of check to time of use* (TOCTOU, see also Sect. 3.2).

The approach described by Goldman et al. [53] also aims to link configuration data to the endpoints of a secure channel. The authors work with the predecessor of TLS, the *Secure Socket Layer* (SSL [50]) protocol. The basic idea is to add a measurement of the SSL certificate to the integrity measurement list that is derived from executable code. The authors do not state how exactly they prevent MitM attacks. The SSL certificate could originate from another platform, or, in the case of our attack scenario, from the NIC that smuggles in the certificate via DMA. Furthermore, the NIC can attack the endpoint during runtime to compromise data and cryptographic keys. McCune et al. [91] use a similar protocol as introduced

---

[4] See http://www.trustedcomputinggroup.org/ [accessed 25 February 2014].

[5] See Footnote 4.

by Goldman et al. [53]. The authors utilize modern chipset features such as Intel's *Trusted eXecution Technology* (TXT [see 54]) to drastically minimize the size of the TCB. In their adversary model the authors do explicitly allow DMA attacks. The reason is that they propose a security architecture that benefits of an isolated execution environment. When code is executed in that environment interrupts and DMA are turned off. Furthermore, the state of the host processor is required to be saved and restored every time the isolated environment is used. This results in a performance loss. Hence, this approach is only suitable for quick secure operations. Protecting user input entered via a USB keyboard is not possible at all, since DMA is required to copy the keystroke codes from the keyboard to the main memory.

Dietrich [39, 40] also proposes a trusted channel concept based on a TPM as well as on TLS. He aims at reporting platform configuration changes during a session with a remote platform. His approach requires modifications to the TPM. It is uncertain if such hardware modifications are enforceable in practice. Cheng et al. [29] also aim to prevent MitM attacks by combining the TCG-based platform configuration reporting approach with a TLS channel. Unfortunately, the authors do not present an implementation. They also do not clearly describe which TLS handshake message they use for the negotiation of the proposed channel. The approach described by Yu et al. [151] also combines TPM-based platform configuration data with the TLS protocol. The authors strongly focus on the TLS renegotiation attack [see 102]. They claim that this attack is also possible with a trusted channel. We doubt this since the attack protocol flow presented by Yu et al. [151, p. 3] demonstrates that the MitM is required to send authentic platform configuration data to the server. Hence, the server is able to detect the code responsible to conduct the renegotiation attack. The authors do not explain if it is possible for the MitM to forge trustworthy platform configuration data.

Quite an interesting trusted channel approach with regard to privacy was presented by Cesena et al. [27]. The proposed channel is a combination of the *Direct Anonymous Attestation* (DAA [20]) protocol as adapted by the TCG and TLS. In this context, DAA allows a platform to prove that it contains a TPM without revealing which particular TPM it is. This helps to preserve privacy if it is required to avoid the linkage of different sessions to a TPM of a particular platform. Besides DAA, the proposed channel is quite similar to our trusted channel. The authors exploit the TLS handshake messages in a similar fashion to our solution.

Sadeghi and Schulz [109] enhance the secure channel protocol IPsec to implement a trusted channel. The approach uses the *Internet Key Exchange Protocol Version 2* (IKEv2 [75]) as basis to bind platform configuration information to the channel. Configuration data can also be transmitted during an IPsec session. The authors also consider how their approach can be integrated into the TNC architecture. Although the presented approach is backwards compatible, minimal modifications to IKEv2 are required to fully benefit from the trusted channel.

Platform configuration information can also be included in the Diffie-Hellman (DH) key exchange as demonstrated by Stumpf et al. [128]. The authors rely on a command (TPM_Quote) that is responsible to get a signed report of the integrity measurements stored inside the TPM. The DH approach does not mitigate the deficiencies caused by load time measurements.

Lyle and Martin [86] introduced a channel that considers web service technologies. Their special environment does not allow to apply a TLS-based channel. They combine the TCG platform configuration reporting approach with so-called message-level cryptography [see 86, p. 4]. Chang et al. [28] merged the TCG approach with the *Secure Real-time Transport Protocol* (SRTP [12])/*Z Real-time Transport Protocol* (ZRTP [154]). SRTP/ZRTP provides a secure channel for *Voice-over-IP* (VoIP [34]) transmissions. The authors aim to provide a trusted channel with the combination of the TCG approach and SRTP/ZRTP.

Unfortunately, all the presented TPM-based approaches do not consider runtime attacks (especially DMA-based runtime attacks) sufficiently. They also suffer from the deficiencies described in the beginning of Sect. 3.2. Please note, our work on trusted channels was also originally based on the TPM. In this work, we adapt the trusted channel concept for another attack scenario where attacks originate from peripherals. Our trust model (see Sect. 2.7) considers a different TCB. We do not count on load-time integrity measurements. A TPM is not required. Our measurements are based on a runtime monitor that derives state information based on memory bus transactions, see Chap. 5. These measurements are considered by the secure communication channel that we use in this work. The channel is based on the trusted channel concept that we [10, 52] introduced in prior work.

3.3.2 Co-processor and Smart Card Based Approaches

The approaches described by Jiang et al. [74] and Chess et al. [30] are based on a secure co-processor. The co-processors are used to establish trust by implementing a concept called trusted co-servers. The co-servers execute evaluated and certified programs to authenticate the main servers to be able to monitor their behavior. The co-servers are more secure against physical manipulation. However, they are more expensive than off-the-shelf hardware. Such co-servers are usually implemented in the form of PCI(e) cards with a dedicated processor, RAM, DMA engine as well as ethernet connectors [72, 73]. As such, they are a perfect host for DMA-based malware.

The trusted channel protocol proposed by Akram et al. [5] is intended for a special smart card scenario. The focus of the authors is on a privacy preserving protocol for the smart card user. Hence, their approach is only applicable in scenarios that involve a smart card whereby the identity of the user must not be revealed. In an earlier publication Akram et al. [4] presented another channel that is intended for runtime authentication and verification of a smart card application. That channel is part of a framework that the authors implemented. The channel protocol was also verified by the authors. The channel described by Akram et al. [4] also focuses on smart cards scenarios. Another approach that is based on smart cards was presented by Wang et al. [143]. The authors propose and formally verify a trusted authentication protocol for *Digital Rights Management* (DRM [3]) scenarios. The protocol also considers platform configuration values. The authors do not evaluate an implementation of their proposed protocol.

Chapter 4
Study of a Stealthy, Direct Memory Access Based Malicious Software

In God We Trust; All Others We Monitor.
Motto of the Air Force Technical Application Center,
Part of the Air Force Intelligence,
Surveillance and Reconnaissance Agency

The arms race between malware developers and the anti-malware community reached a new level. Countermeasures for kernel level [60], hypervisor-based [77], and system management mode based malware [49] were proposed [25, 51, 107]. As a result researchers explored new environments for stealthy malicious software. Malware can be placed on dedicated hardware such as video cards and network interface cards to attack the host platform [see 47, 134, 135]. Such devices bring, among other things, a dedicated processor and dedicated runtime memory. These devices can operate independently from the host system. Anti-virus software cannot detect malicious code stored in separate memory and executed on a different processor. An attacker can use such devices, or more precisely, the direct memory access mechanism to circumvent protection mechanisms built into the operating system by attacking the host runtime memory directly. We call code performing targeted DMA-based stealthy attacks to locate and read or modify target data *DMA malware*. Such data can be cryptographic keys for encrypted harddisks, credentials for online banking accounts, instant messenger chat sessions, and open documents located in the file cache.

In this chapter we characterize DMA attacks and derive the term DMA malware. We explore the term by examining if DMA malware can significantly increase the probability of performing a successful stealthy attack against a computer platform while preserving efficiency and effectiveness. For the evaluation we built our DMA malware DAGGER—a DmA-based keystroke loGGER that exfiltrates captured data to an external entity. We are interested in the efficiency, effectiveness and especially in the stealth properties of DMA malware. We chose to implement a keystroke logger to demonstrate that "short living" data can be captured by DMA malware.

Our implementation is based on Intel's manageability engine that is part of the popular x86 platform. Intel's ME is implemented in business as well as consumer platforms (see Intel vPro platforms [66]) to support different applications, such as the *Intel Active Management Technology* (iAMT [39]) or the *Identity Protection Technology* (IPT [67]). Our DMA malware DAGGER is not executed on the host

© Springer International Publishing Switzerland 2015
P. Stewin, *Detecting Peripheral-based Attacks on the Host Memory*,
T-Labs Series in Telecommunication Services, DOI 10.1007/978-3-319-13515-1_4

processor. It is executed on the processor provided by Intel's ME. No additional hardware is required. DAGGER implements an isolated runtime attack on user input. Additionally, our DMA malware could steal cryptographic keys, target OS kernel structures in an attack, and copy files from the file cache. Although DMA malware cannot by detected by anti-virus software, an attacker still faces certain challenges. DMA malware must be effective, i.e., it should be able to successfully attack various systems. DMA malware must also be efficient, i.e., fast enough to find and process data, even when dealing with virtual memory addresses and randomly placed data. Such malware goes beyond the capability to exploit DMA hardware.

The main contributions of this chapter are:

- **DMA malware definition**: There are different kinds of code that utilizes DMA. To clearly identify if code should be considered harmless, an attack, or DMA malware, we introduce an appropriate definition.
- **DMA malware core functionality**: We present a number of requirements that must be fulfilled by DMA malware in order to mount successful attacks.
- **Evaluation of DMA malware prototype implementations**: To demonstrate that DMA malware increases the probability for successful stealthy attacks while preserving efficiency and effectiveness, we implemented DAGGER. DAGGER is executed on Intel's isolated ME. DAGGER operates stealthily and can attack multiple operating systems. Our implementation is fast and efficient that it can capture keystrokes very early in the platform boot process, that enables DAGGER to capture harddisk encryption passwords under Linux, for example.
- **DMA side effect detection approach**: We present a detection approach that can reveal DMA malware executed in isolated hardware environments. Our work demonstrates that DMA malware produces unexpected side effects that we measure utilizing widely used and cross platform available CPU features.

4.1 DMA Malware Definition

To define the term DMA malware we first characterize different kinds of DMA-based code. This helps to clearly distinguish between simple DMA usage, DMA attacks and DMA malware, whereby the latter has a clear focus on stealthiness. Note, DMA malware goes beyond the capability of controlling a DMA engine. DMA-based code that implements malicious functionality is considered a serious threat. Such code can be operating stealthily during infiltration and runtime. It is also an advantage, e.g., for long-term attacks, if the code can survive platform reboots and power off as well as standby modes. Hence, we can prioritize the following criteria to assess code that utilizes DMA. That is, the DMA-based code:

(C1) implements malware functionality
(C2) needs no physical access to increase the probability of stealthy infiltration
(C3) applies rootkit/stealth capabilities during runtime
(C4) can survive reboot/standby/power off modes

Table 4.1 Fulfillment of criteria C1–C4 of DMA attack examples

Attack presented in	C1	C2	C3	C4	DMA malware
[90] (USB)	–	–	–	✓	–
[15, 17–19, 42, 43, 87, 101] (FireWire)	✓	–	✓	✓	–
[11, 61, 87] (PC card)	✓	–	✓	✓	–
[131] (Intel ME)	–	✓	–	✓	–
[35, 36, 47] (NIC)	✓	✓	✓	✓	✓
[134, 135] (Video card and NIC)	✓	✓	✓	✓	✓
[80] (Video card)	✓	✓	–	–	–

Note, the assessment was done using publicly available material. If we could not decide with the help of available resources whether a criterion is fulfilled, we assume that this criterion is fulfilled.

We use a binary system for our prioritization:

$$2^3 \quad 2^2 \quad 2^1 \quad 2^0$$
$$\text{C1} \quad \text{C2} \quad \text{C3} \quad \text{C4}$$

This system distinguishes 16 kinds of DMA-based code. We can derive a unique number for each kind. For example, DMA-based code that does not perform malicious actions ($\text{C1} = 0$), leaves no traces on the host ($\text{C3} = 1$), does not need physical access ($\text{C2} = 1$), and cannot survive reboots ($\text{C4} = 0$) is mapped to the binary pattern 0110. This pattern corresponds to class 6 in decimal. The higher the derived number, the more dangerous is the DMA-based code.

Our definition of DMA malware is as follows:

Definition: DMA malware is malicious software executed on dedicated hardware attacking a computer system via a mechanism called direct memory access as well as fulfilling at least the criteria C1, C2, and C3.

When applied to the target platform introduced in Chap. 2, this definition means, that DMA malware is based on first-party DMA and the DMA engine can be configured by the attack code to not involve the host CPU. The attack code is executed on dedicated hardware with its own processor and runtime memory, such as a NIC. Controlling the NIC increases the probability that an attacker can hide data during exfiltration. Table 4.1 applies our binary system to the DMA attacks that are presented in Chap. 3 "Related Work". The table also depicts what related work is DMA malware according to our definition. In this chapter we also aim to develop a DMA malware proof of concept that fulfills at least the criteria C1, C2, and C3.

4.2 DMA Malware Core Functionality

When attacking the host, it is not enough for an attacker to control a DMA engine. The engine enables the attacker to read from and to write to host memory. However, in most cases the target memory address is not known. This section describes the core functionality of DMA malware, i.e., overcoming address randomization, memory mapping, and search space restriction.

The attacker has to determine memory addresses. The problem is that the memory space allocated for, e.g., kernel data structures is not at the same memory address after a platform reboot. Data structures are placed *randomly in memory* by the OS. This can happen in a natural way when a device driver, for example, allocates memory and gets the next free unallocated memory chunk. The memory address of that chunk is not necessarily the same after a platform reboot. Alternatively, the OS can apply certain randomization algorithms to ensure that data structures are not placed at the same memory position. Of course, an attacker can scan the whole system memory for signatures of the target data, but this is very inefficient when scanning a system with 4 GB physical memory or more.

Operating systems work with *virtual memory addresses* [see 31, Chap. 15]), but DMA works with *physical memory addresses*. The OS creates so-called page tables that are used by the host CPU to map virtual memory addresses to physical ones. The mapping is absolutely necessary to resolve memory address pointers when using DMA. A special host processor control register called CR3 contains the physical memory address of the page tables. The attacker has no access to the CR3 register. The visibility of a DMA engine is restricted to host memory only. Without further analysis the attacker has to scan the whole memory address space for relevant data. There are two potential ways in which an attacker can overcome this problem. The first way is to analyze if the OS places the data structures in question in approximately the same memory area. The second possibility is to implement OS memory management mechanisms. That is, the attacker must find a way to access memory page tables created by the OS. With access to the page tables the attacker can then traverse page tables and is able to resolve pointers from one data structure to another. This still requires a known starting point for the search.

4.3 Design and Implementation of DAGGER

We present an overview of a general design for our DmA-based keystroke loGGER DAGGER in the next subsection before we explain the details of the DAGGER implementation in Sect. 4.3.2.

4.3.1 General Design

Our design of DAGGER is depicted in Fig. 4.1. DAGGER is DMA malware. That is, DAGGER has to fulfill the DMA malware definition including at least the criteria C1, C2, and C3. DAGGER consists of three main components:

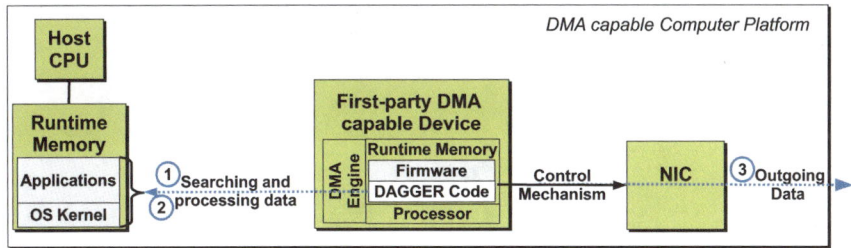

Fig. 4.1 General design of DAGGER. DAGGER is executed on a DMA capable device so that it can (*1*) search and (*2*) process data from host runtime memory. It controls a communication path to exfiltrate information (*3*)

- **Search:** find the address of valuable data in the host memory via DMA.
- **Process data:** read valuable data within the regions identified during the search process.
- **Exfiltration:** exfiltrate information in a way that is invisible to the host.

4.3.2 Implementation Based on Intel's ME Environment

To evaluate DMA malware we chose to implement DAGGER on Intel's ME. Intel's ME provides some useful features for implementing DMA malware that we describe in the following.

The core of Intel's ME is an embedded micro-controller placed in the platform's MCH. This isolated environment contains *Read Only Memory* (ROM), *Static Random Access Memory* (SRAM), DMA hardware to access the host memory [25, 131], and a processor as depicted in Fig. 4.2. The embedded processor of the ME is an ARCtangent-A4 (ARC4). The isolated environment is available regardless of the power state, even in standby or power on/off. It only requires that the chipset is connected with a power source. Applications executed on the embedded micro-controller are implemented in firmware (ME FW) and stored in flash memory together with the BIOS. The most prominent ME firmware example is Intel's Active Management Technology. But depending on the kind of computer platform (business or consumer hardware) the ME can also run other firmware. Other firmware executed by Intel's ME are for instance: Intel's Identity Protection Technology, *Alert Standard Format* [131, p. 46]), *Intel Quiet System Technology* (QST [131, p. 46]) for temperature and fan control, and *Integrated Trusted Platform Module* (iTPM [79, p. 109]).

ME firmware can communicate with the host via a PCI device called *ME Interface* (MEI [79, p. 71]). The MEI can provide the version of the executed ME firmware, for example. The ME environment provides additional PCI devices[1] to support certain

---

[1] These devices can act as bus masters, see Sect. 2.5.

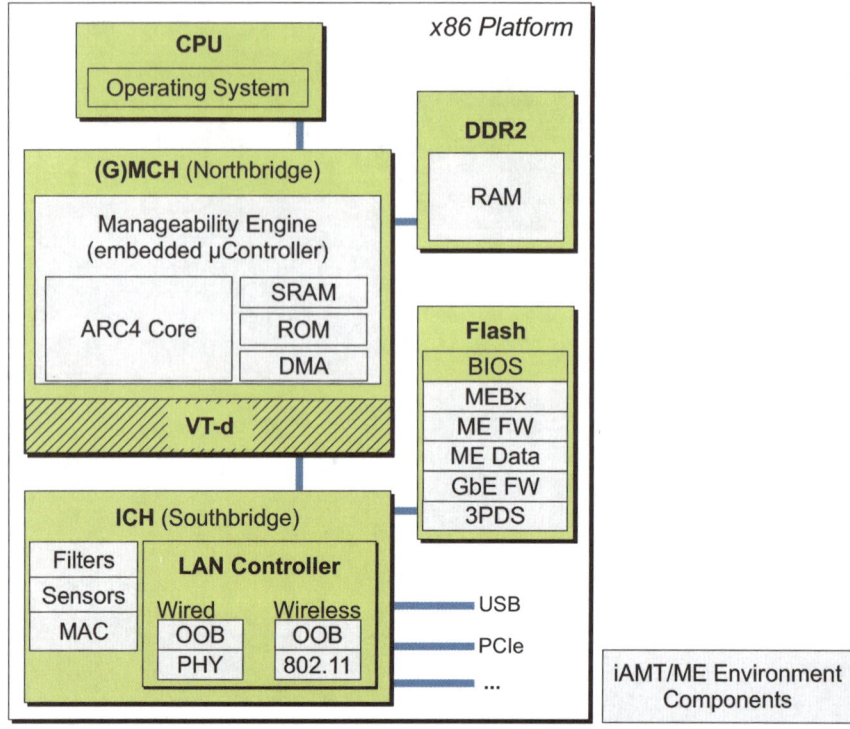

Fig. 4.2 Intel's Manageability Engine environment. Intel's Manageability Engine (ME) environment consists of the Manageability Engine that is included in the MCH. Furthermore the environment consists of an isolated part of the RAM as well as isolated portions of persistent flash memory. The ICH also contains ME environment components, especially components that implement the out-of-band channel

AMT features such as text console and disk redirection. A serial port is emulated to implement text console redirection [see 79, Chap. 5]. Text output that is sent to this port is forwarded to a remote console via the network. With this capability an administrator can remotely control the BIOS. To implement disk redirection a local disk is emulated by the ME environment [see 79, Chap. 5]. An administrator can remotely mount storage media (e.g., a CDROM with an operating system installer to recover the operating system of the AMT enabled platform) via the locally emulated disk.

During the platform power-on procedure the ME firmware image is loaded into ME RAM. The ME firmware itself runs on the micro-controller internal ARC4 processor and it also uses some system RAM as depicted in Fig. 4.2 to store runtime data. This runtime storage is provided by a certain memory area that is invisible to the main CPU and the OS. The separation is enforced by the chipset [79].

The ME environment introduces *Out-Of-Band* (OOB) communication, i.e., a special network traffic channel used by iAMT. The iAMT enabled computer platform is managed by a remote management console using OOB. OOB is also available regardless of the power state. OOB can be considered to be a separate network connection, running on the same hardware. The ICH implements necessary components to support the ME environment with the OOB feature. The firmware filters network traffic intended for, e.g., iAMT and redirects the packets to the ME. The host is unaware of the redirected ME network traffic. This kind of traffic is identified by TCP port numbers.

4.3.3 Attack Implementation Details for Linux and Windows Targets

We implemented two keystroke logger prototypes to attack two targets, Linux and Windows based OSes. We decided to find and monitor the keyboard buffer address of 32 bit versions of the target OSes. In comparison to 64 bit versions, 32 bit versions have to deal with a more complicated memory management. For example, the attacker has to consider *Physical Address Extensions* (PAE [105, p. 769]) or certain memory offsets when mapping memory addresses. The following subsections describe, how we implemented the DMA malware core functionality as described in Sect. 4.2. The prototypes capture short living keystroke codes within their *monitoring phase*. Each prototype handles the *search phase* for the target buffer differently. This has at least two reasons. One reason is to evaluate as many aspects as possible of DMA malware. The other reason is that OSes have different memory management properties. We use a vulnerability described by Tereshkin and Wojtczuk [131] to infiltrate the ME environment during runtime. To call our code we hook a ME firmware function that we identified as the library function memset. Tereshkin and Wojtczuk [131] assumed that they hooked a timer interrupt handler, but they actually hooked the ME firmware function memcpy. We hook memset since we determined that it is called more often.

Our Linux variant is based on a signature scan as depicted in Fig. 4.3. We analyzed the available Linux source code to derive a signature of our target, the physical address

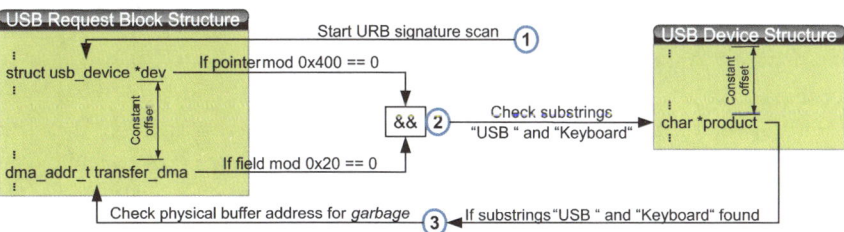

Fig. 4.3 USB request block signature scan (simplified). The scan (*1*) begins to search for a pointer to the USB device structure. A candidate for such a pointer is aligned to a 0x400 boundary. The value of the structure field transfer_dma must be aligned to a 0x20 boundary. If both conditions are true, the product string in the USB device structure is (*2*) checked for the substrings "USB" and "Keyboard" In the last step the signature scan (*3*) checks if the keyboard buffer contains *garbage*, that is, invalid keystroke codes

of the keyboard buffer. The buffer address is part of the *USB Request Block* (URB) structure that is defined in the file `include/linux/usb.h` of the Linux source code. The demanded structure field is called `transfer_dma`. The memory offsets differ from kernel version to kernel version. We solved that problem by exploiting the *Grand Unified Bootloader* (GRUB) that places an identifier at a constant physical memory address. We implemented a function that reads the identifier via DMA and parses the kernel version number to derive corresponding offsets. Afterwards our prototype runs through the search phase, that is, the signature scan.

Since our Linux prototype targets kernel data structures we can restrict the search space to the first gigabyte of system RAM. Standard Linux systems have a memory split of 1 GB/3 GB, that means, 1 GB for kernel space and 3 GB for user space. We were able to further restrict the search space by empirically analyzing in which memory area the kernel places the data structures needed by our signature scan. We determined that this memory area is between `0x33000000` and `0x36000000` for the Ubuntu Linux kernel version 3.0.0 after a fresh platform boot. The address of the keyboard buffer does not change after standby or hibernate mode. With this approach we overcome the problem of inefficiently scanning the whole system memory for the randomly placed signature. Mapping virtual addresses to physical ones is a minor issue when attacking the Linux kernel. Normally, in 32 bit versions a kernel virtual address (or more precisely kernel logical address [see 31, Chap. 15]) is mapped to its physical address by subtracting a constant offset. In 64 bit Linux versions such an offset is not needed. Hence, there is no need to know the content of the `CR3` processor register.

The search strategy for Windows-based target platforms works different. To be able to perform the search using the search path as described below, virtual addresses must be mapped to physical ones. This mapping is done using page tables created by the Windows kernel. The memory address of those page tables is loaded into the `CR3` register, which an attacker cannot access via DMA. It turned out after some empirical tests with a simple driver, that the physical address of the page tables for the *system process* takes one of the following two values for Windows Vista/7 systems: `0x122000` or `0x185000`. The system process is the first process created during Windows startup. With this knowledge DAGGER can access the page tables created by the kernel and overcomes the problem of mapping virtual addresses to physical ones. DAGGER implements a page table traversing algorithm that takes account of PAE.

Our Windows malware searches for a structure called `DeviceExtension` that is maintained by the USB keyboard driver `kbdhid.sys`. This structure contains a buffer that stores the codes of the last pressed keys. The source code for `kbdhid.sys` is not publicly available. The most convenient way to get internal information of that driver was to use *IDA Pro*,[2] *Windows Debugger* (WinDbg) tools, and debug symbols provided by Microsoft[3] in the form of `pdb` files. To finally

---

[2] See http://www.hex-rays.com/products/ida/index.shtml [accessed 25 February 2014].

[3] See http://msdn.microsoft.com/en-us/windows/hardware/gg462988 [accessed 25 February 2014].

determine the location of the buffer in the DeviceExtension structure, our research starts early in the boot process [see 105, Chap. 13]. We analyzed further internal Windows structures. To find a starting point for the search, we analyzed the *Kernel Processor Control Region* (KPCR [105, p. 62ff]), or more precisely KiInitialPCR, the KPCR for the processor 0. We also examined the *Object Manager Namespace Directory* (OMND, part of the Windows object manager). We determined that KiInitialPCR is well suited to derive a path to the DeviceExtension structure as depicted in Fig. 4.4. KiInitialPCR is not located at a constant memory address. DAGGER has to apply another step before it can start with the search as depicted in Fig. 4.4.

The memory position of KiInitialPCR is determined by a function called OslpLoadAllModules of the winload.exe binary as depicted in Fig. 4.5. This binary is loaded by the Windows boot manager bootmgr that in turn is loaded by *Master Boot Record* (MBR) code, etc. The function loads the *Hardware Abstraction Layer* (HAL) library hal.dll as well as the Windows kernel image in a more or less random manner. The kernel image contains KiInitialPCR at a constant

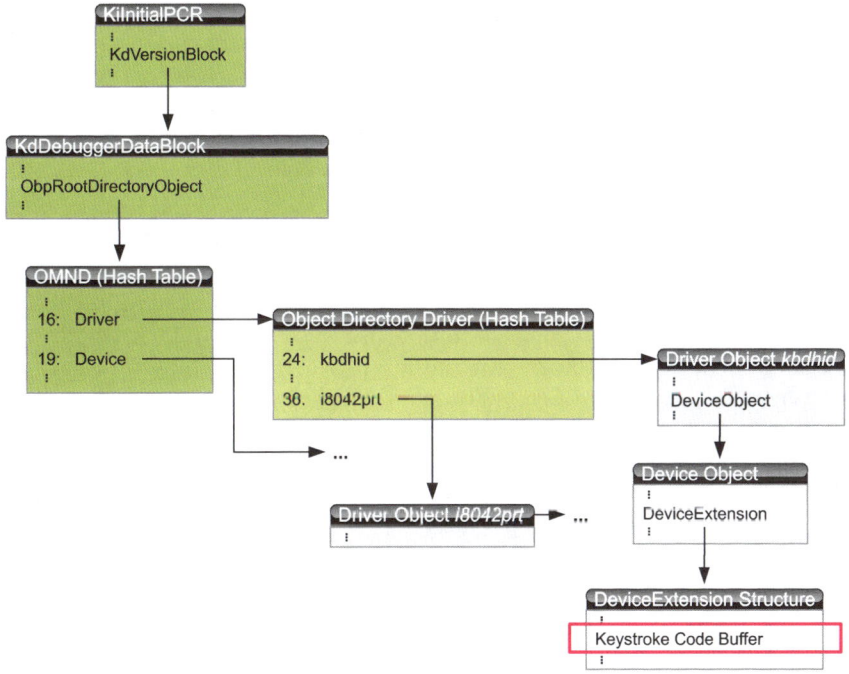

Fig. 4.4 Find DeviceExtension structure (simplified). With KiInitialPCR as a starting point, DAGGER finds the OMND, that provides via hash tables a path to the driver object kbdhid. This object contains a pointer to a device object. The device object provides the DeviceExtension structure, which contains the keystroke code buffer

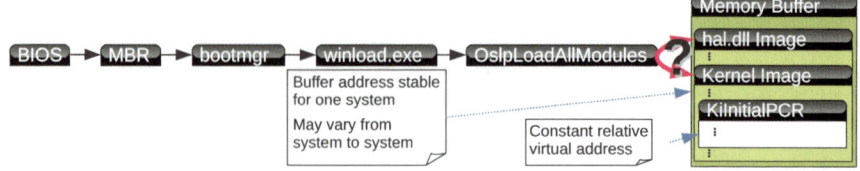

Fig. 4.5 Find `KiInitialPCR` (simplified). `OslpLoadAllModules` determines the exact position of the Windows kernel image and the HAL

relative address. The disassembled code of `OslpLoadAllModules` is similar to an *Address Space Layout Randomization* (ASLR [105, p. 757]) mechanism.

The memory buffer for the kernel image and the HAL is allocated by `Oslp-LoadAllModules` via a function called `BlImgAllocateImageBuffer`. The latter function returns stable address values for a Windows system. These values may vary on different systems. For every possible return value of the function `BlImgAllocateImageBuffer` there are 64 theoretically possible different 4 KB aligned virtual addresses. These addresses need to be checked in order to find the kernel image base address. The disassembly of `OslpLoadAllModules` revealed that the randomization seed for the address randomization has a 5 bit value. This implies 32 possible addresses for each (of two) possible load order cases, i.e., first kernel image and then `hal.dll` or vice versa. As long as `KiInitialPCR` has a constant relative virtual address within the kernel image, the same number of virtual addresses to be checked also applies for a direct `KiInitialPCR` search without any need to deal with the kernel image. To ensure that DAGGER found the correct `KiInitialPCR` we implemented a `KiInitialPCR` signature check. When DAGGER has identified the correct `KiInitialPCR`, it continues to look for the keyboard buffer using the search path described in Fig. 4.4.

We use ethernet controller to exfiltrate the captured keystroke codes. To be more precise, we use the OOB features of the Intel ME environment. Unfortunately, there is no documentation that explains how to use this feature. Hence, we had to analyze the firmware to figure out how to exfiltrate keystroke codes using the OOB channel. We were able to find the transmit ring buffer that is used to send network packets in the ME runtime memory. Furthermore, we were also able to find the firmware code that is responsible for sending the next network packet from the transmit ring buffer. To exfiltrate the captured data we prepare network packets, e.g., DHCP discover packets as depicted in Fig. 4.6, that contain the logged keystroke code. Then, we copy the prepared network packet to the transmit buffer. Afterwards, we trigger sending the packet by the NIC to an external platform. Please note, the transmitted packets can easily be found when analyzing the network traffic with an external platform. To improve the stealthiness of the design we [124, 125] implemented a covert timing channel that is based on a so-called Jitterbug [see 115].

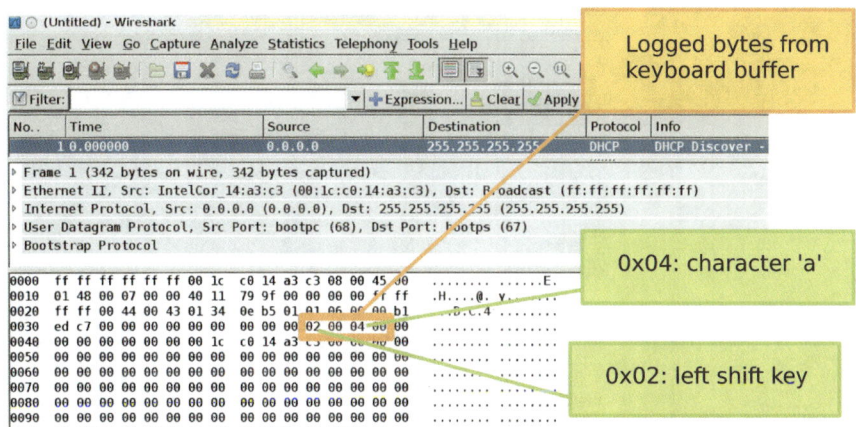

Fig. 4.6 Network packet containing bytes from keyboard buffer. The wireshark instance is executed on an external platform. The network packet that has been parsed by wireshark contains 4 bytes that represent the logged keystroke code data

4.4 Evaluation

We used an x86 platform with a Q35 chipset, 2 GB RAM, a 4-core 3 GHz CPU, and iAMT firmware (version 3.2.1) to evaluate DAGGER with four different 32 bit OS kernels: Windows Vista Business (Service Pack 2), Windows 7 Professional (Service Pack 1) and Ubuntu Linux kernel version 2.6.32 as well as kernel version 3.0.0.

4.4.1 DMA Malware Fulfillment

We designed and implemented our DAGGER prototypes according to the DMA malware definition described in Sect. 4.1. (C1) is clearly fulfilled since DAGGER implements working keystroke logger functionality. DAGGER needs no physical access for the infiltration process (C2). We infiltrate the ME environment using a software-based exploit during runtime. DAGGER exploits dedicated hardware to implement rootkit properties (C3). We ran host performance overhead tests (memory: MEM, network: NET, and CPU), since host and ME environment share the NIC as well as a RAM chip. Parallel NIC and RAM accesses must be arbitrated and could therefore cause delays. Our measurement results depicted in Fig. 4.7 reveal no significant overhead. The highest overhead that we could detect is approximately 1.5 % when accessing the host memory during the search phase. It is extremely unlikely that this minimal overhead would reveal DAGGER.

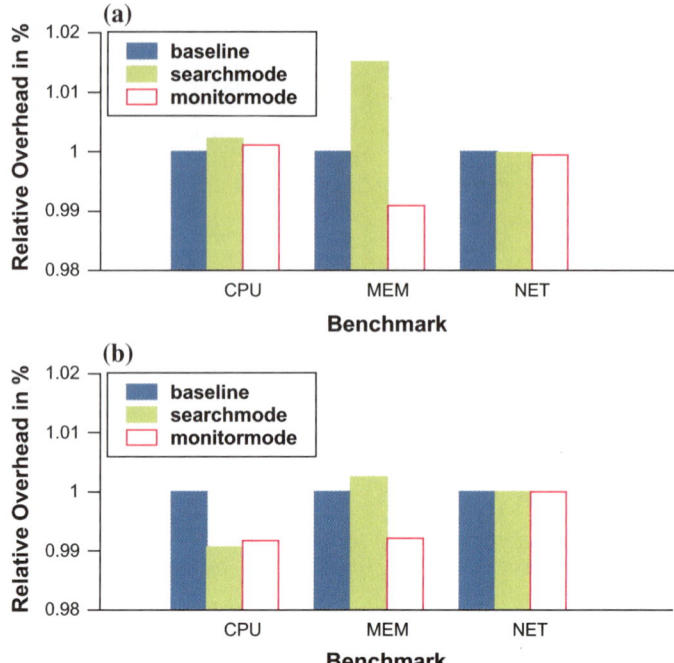

Fig. 4.7 Host performance CPU, MEM, and NET overhead tests. **a** Linux 3.0.0 performance overhead test results. **b** Windows 7 performance overhead test results. We used time stamp counters to measure overhead time. We measured the time it takes to copy a 100 MB test file over the network (NET) and within RAM (MEM) as well as the time it needs to compute a SHA1 hash sum over this test file ten times in parallel to stress all four CPU cores (CPU). Each benchmark was performed three times: without keystroke logger (*baseline*), keystroke logger in search mode, and keystroke logger in monitoring mode. For the monitoring mode we configured the keystroke logger to constantly send network packets of approximately 1,000 packets per minute. This is equal to 500 keystroke and 500 key release events. We repeated each test 1,000 times. A bar in the gure represents the mean of 1,000 runs

The search times summarized in Fig. 4.8 are very short and the very aggressive memory stress test we performed does not represent the memory utilization of a normal computer system. DAGGER has solely read-only operations to ensure stealthiness. The popular network sniffer *Wireshark*[4] was not able to detect any DAGGER traffic on Linux and Windows systems. Host firewalls cannot block such traffic either. Even if anti-virus software knew DAGGER's signature it would be unable to access DAGGER's memory to apply the signature scan successfully. Nevertheless, we also run a software called *Mamutu*,[5] that is, amongst other things, specialized in detecting keylogger behavior. Even specialized software could not find any indication of DAGGER. Regarding criterion **C4** we successfully checked if

---

[4] See http://www.wireshark.org/ [accessed 25 February 2014].

[5] See http://www.emsisoft.com/en/software/mamutu/ [accessed 25 February 2014].

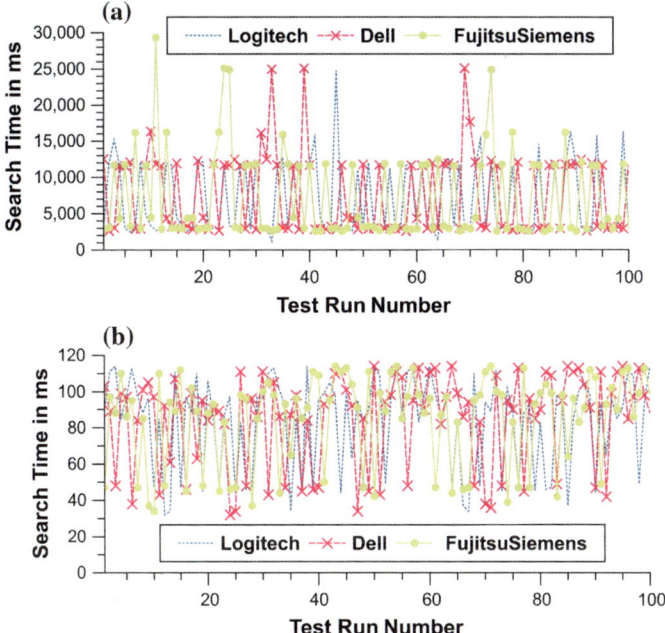

Fig. 4.8 Search time measurement results. **a** Linux 3.0.0 several keyboards **b** Windows 7 several keyboards. The test results with several keyboards under Linux reveal a best case for search times of around 1,000 ms and a worst case of almost 30,000 ms as depicted in (**a**). The median for all keyboards is at 3,281 ms. Useful for comparison: scanning the whole memory area determined for Linux (see Sect. 4.3.2) search takes approximately 13,000 ms. The worst case of 30,000 ms is due to an erroneous DMA transfer that we do not handle directly. This causes DAGGER to repeat the search phase. On Windows 7 the best search time is approximately 50 ms and the worst time is around 120 ms, see (**b**). The median for all keyboards is at 93 ms. Hence, the search strategy we implemented for Windows targets performs much better than the signature scan based strategy for Linux

DAGGER's attack code is fully functional after a platform reboot, after standby and after power off state. We determined that this depends on an iAMT BIOS option. Our code cannot survive a cold boot that happens if this option is not set.

4.4.2 Effectiveness and Efficiency

DAGGER is efficient, since it can permanently catch short living data from the keyboard buffer. To demonstrate that DAGGER is also effective we tested DAGGER with different Windows and Linux versions as well as several keyboards. The measured search times summarized in Fig. 4.8 confirm that DAGGER is quite efficient. We repeated the measurements for each kernel and for each keyboard 100 times. We took a measurement after a platform (re)boot to change the target address for each test run. The Linux measurement results imply that we could further restrict the search space. We could start the search near the lowest address we encountered most often during our tests. Search times of around 2,500 ms are due to target addresses near 0x33c00000. Thus, we could skip almost 2,500 ms if we start the search at 0x33c00000. Furthermore, we could skip the search area address range between 0x34000000 and 0x36000000. Almost no targets were found in this area. A lot of targets were found near 0x36e0000, i.e., search times of around 12,500 ms that could also be saved. This increases the probability to miss keyboard buffer addresses. That is, we can get better search times at the expense of effectiveness. The best case search times are sufficient to capture hard disk encryption passwords, for example. We tested this successfully with a Linux system. The Windows kernel can swap out memory pages to the hard disk—Linux does not. Swapped memory pages cannot be found by DMA malware. Hence, we also did a test for Windows to check if swapping has any effect on DAGGER as depicted in Fig. 4.9b.

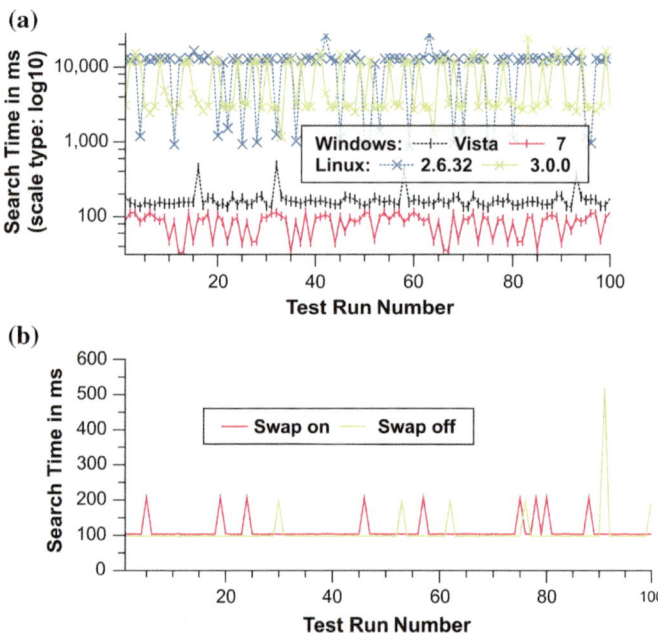

Fig. 4.9 Search time measurement results. **a** Several operating systems. **b** Windows 7 swap on/off. The plot in (**a**) compares different target kernels. DAGGER performs slightly better on Windows 7 than on Windows Vista. Linux 2.6.32 places the target memory structure closer to 0x33000000 than Linux 3.0.0. Thus, DAGGER has more hits around 1,000 ms when attacking Linux 2.6.32. The results in (**b**) confirm that swapping has no effect on the efficiency and effectiveness of DAGGER. A platform reboot was only applied to change the swapping behavior. The peaks are due to restarts of the search phase

4.4.3 ME Firmware Condition

To be really stealthy DAGGER ensures that the ME firmware is still up and running correctly. iAMT provides a web server for remote platform management [see 79, p. 215] that is still usable. The server responds correctly on the local platform on Linux and Windows. Firmware tools utilizing the MEI (see Sect. 4.3.2) also work when DAGGER is active. We successfully tested the *AMT Status Tool* (part of the *Local Manageability Service* driver) and the *Manageability Connector Tool* (part of the *Manageability Developer Toolkit 7.0*) under Windows. Under Linux we successfully tested the *Intel AMT Open-source Tools and Drivers* (version 5.0.0.30), or more precisely the *ME Status* and the *ZTCLocalAgent* tool. Note, we determined that DAGGER still runs even after having disabled the iAMT firmware in the BIOS. It appears that the ME environment cannot be disabled entirely via any BIOS options.

4.4.4 I/OMMU

To test an I/OMMU (see Sect. 2.6) as a countermeasure against DAGGER we enabled Intel VT-d in the BIOS. As far as we know Windows does not support I/OMMUs directly. We could successfully attack Windows Vista and Windows 7 although the I/OMMU was activated. Linux supports I/OMMU configuration with additional effort. We also enabled VT-d in the BIOS and we activated I/OMMU support via the kernel command line. With these additional steps we were able to prevent the Linux version of DAGGER from reading short living keystroke codes from OS memory. This protection is not activated by default. In the next section we discuss, among other things, further issues regarding the I/OMMU.

4.5 Countermeasures Considerations

To scan for DMA malware using software executed on the host CPU is quite difficult. For example, current AV software does not scan the runtime memory of peripherals or the host CPU cannot access the runtime memory due to certain isolation mechanisms. The worst case for a scanning approach is that the DMA malware changed the behavior of the scan software, which would deliver incorrect results. Checking firmware images at load time, as proposed by the TCG [136], does not prevent runtime attacks. Furthermore, it is unclear if all ROM components are accessible by the host.

4.5.1 I/OMMU Issues

In the case of DMA attacks an appropriate configuration of the I/OMMU (see Sect. 2.6) is proposed as a preventive countermeasure, for example by Duflot et al. [47, p. 48]. It is required that system software configures the I/OMMU. An incorrect configuration cannot be excluded [83, p. 2].

It is assumed that the I/OMMU is secure. Unfortunately this is not always the case. Sang et al. [111] demonstrated that an I/OMMU configuration can be tricked with legacy PCI devices. Wojtczuk et al. [148] revealed that an I/OMMU can be attacked by modifying the number of DMA remapping engines provided by the BIOS (see Sect. 2.6). This is done before the I/OMMU is configured by system software. The environment we used for DAGGER is able to carry out such an attack. This threat can only be mitigated by executing special hardware dependent code called SINIT. However, on at least one previous occasion the manufacturer of the chipset failed to release SINIT code at the launch of the chipset [147, p. 22]. This code is needed to initialize a well known and trustworthy environment for, e.g., a hypervisor. It checks the DMA remapping engines and can therefore prevent an attack as presented Wojtczuk et al. [148].

SINIT belongs to and increases the size of the trusted computing base. Previous work demonstrated that SINIT code can have exploitable security vulnerabilities that can be used to trick I/OMMU mechanisms [see 148]. Recently, Wojtczuk and Rutkowska [148] presented another attack that can be used to circumvent I/OMMU mechanisms as well. To prevent the attacks presented by Wojtczuk and Rutkowska [146, 148], a SINIT as well as a BIOS update must be applied. Wojtczuk et al. [147] presented another I/OMMU attack. Note, SINIT is normally triggered on hypervisor-based platforms. Platforms running a normal OS cannot necessarily count on the I/OMMU. It should also be mentioned that SINIT requires the activation of additional platform features, namely the *Trusted eXecution Technology* and the TPM [54]. This means that users that do not want to activate the TPM for example cannot rely on the I/OMMU. Note, the TPM is an opt-in device [see 54, p. 212] and is turned off by default.

For a comprehensive protection against DMA malware it is absolutely necessary to correctly configure the I/OMMU. However, the I/OMMU can only be considered secure if the above mechanisms to protect the whole platform are secure. This is a difficult task. Hence, alternative approaches were considered by Li et al. [83] and Duflot et al. [46]. Li et al. [83] state that their approach requires extending the firmware, does not work correctly if peripherals cause heavy PCIe traffic, and the verifier component needs to know the exact hardware configuration. The approach presented by Duflot et al. [46] is highly NIC adapter-specific and not applicable to isolated environments such as Intel's ME. It is worth noting that malware such as our implementation controls the NIC without any NIC firmware modifications, i.e., exfiltration cannot be detected by the approach described by Duflot et al. [46]. Furthermore, this approach has significant performance issues for the host CPU (100 % utilization of one CPU core).

Memory access policies enforced by I/OMMUs can be insufficient or can even prevent the use of some other features in some application scenarios. Consider hardware supported malware scanners such as *CoPilot* [100] and *DeepWatch* [25]. The I/OMMU can be configured to stop CoPilot and DeepWatch from working or to allow such systems to access the host memory to scan it for malicious software. In the latter case DMA malware could make use of the execution environment of CoPilot or DeepWatch to attack the host. DAGGER, for example, uses the DeepWatch environment, i.e., Intel's ME. Since iAMT version 5, Intel supports a verified launch for the firmware to be executed on Intel's ME [see 79, p. 271]. The firmware is checked during load time. The result of the load time check is provided to system software. As far as we know the result is not used in practice. The mechanism cannot prevent runtime attacks as applied by our PoC. This means, DAGGER confirms that our assumption that an attacker already infiltrated the target system, e.g., via a zero-day exploit (see Sect. 2.7), can also hold even if such additional security mechanisms are in place. An appropriate configuration of the I/OMMU is a first step against DMA malware. However, without resolving the mentioned issues a successful deployment cannot be guaranteed.

4.5.2 Detection Approach Based on DMA Side Effects

A possible detection approach is based on DMA side effects that we observed in a first experiment with our own DMA malware prototype DAGGER. Our detection mechanism is based on multiple widely used and cross platform CPU features.

So far we developed, implemented, and evaluated our mechanism that is able to detect rogue DMA usage that is not initiated and unexpected for the host system. DMA usage is initiated by the host CPU when a peripheral has to process data on behalf of the host CPU. Sending a network packet using the network interface card is an example. Expected DMA usage originates from peripherals and is intended for software running on the host CPU such as the operating system. Receiving a network packet is an example for intended DMA usage. Our method is able to detect a general side effect pattern. Thus, we believe it is suited to detect other kinds of DMA malware besides the prototype we implemented. Our investigation into detecting malicious DMA usage is based on the knowledge that both, the main CPU and platform peripherals, can request to access the main system memory at the same time. The memory controller hub arbitrates parallel memory access requests, see Fig. 2.5. The interesting question for us was if this parallel memory access introduced any measurable side effects. If side effects are present and measurable then we can use these to detect malicious behavior.

We booted a Linux kernel and started just a root shell to ensure that the system workload was minimized. Only one CPU core was online. We performed a memory stress three times: without keystroke logger (baseline), keystroke logger in search mode, and keystroke logger in monitor mode, see also Sect. 4.3.3. For the tests we used a 100 MB file that we copied from one location to another within a

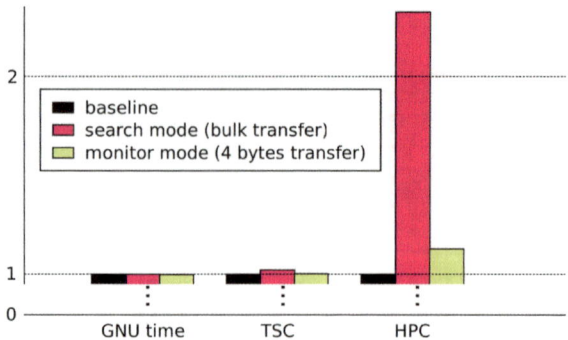

Fig. 4.10 Memory stress measurements. Search phase and monitor phase are depicted relative to the baseline

RAM-based file system. We repeated the tests 1,000 times and calculated the means. The results are depicted in Fig. 4.10. The diagram reveals how we refined our strategy with different and more specialized measurement tools.

GNU Time Measurements First we tried the common system tool GNU time to determine a delay. GNU time measures system resource usages of a process, in our case the memory stress test tool. As shown in Fig. 4.10 on the left hand side the means of the test runs are nearly the same. We concluded that the measurement resolution of GNU time is insufficient to reveal delays in our experiment.

Time Stamp Counter (TSC) Measurements We repeated our measurements with a more accurate hardware-based measurement tool, the TSC [see 69, Sect. 17.12]. The TSC counts clock ticks, see Sect. 2.3. The results are presented in the middle in Fig. 4.10. We were able to (re)produce an overhead of 2 % when our prototype malware is in search mode. DMA was originally introduced to eliminate the burden on the CPU. That means, to perform memory transfers without the involvement of the host CPU. *Hence, that overhead is surprising and a first piece of evidence that detectable DMA side effects exists.* When our prototype malware is in monitor mode we cannot see noteworthy overhead when using TSC. The critical difference between the two modes is that in search mode the malware copies at least a memory page where it searches for valuable data. However, in monitor mode the malware copies just 4 bytes from the keyboard buffer.

Hardware Performance Counter (HPC) Measurements We repeated the measurements with a third approach using HPCs, a hardware-based performance monitoring tool for code optimization, see Sect. 2.3. These counters are special purpose processor registers on Intel processors [69, Chaps. 18/19] that count certain events such as cache misses, branch prediction misses, and resource stalls. Similar HPC are also available on other platforms such as ARM and SPARC. The Intel platform we used for our experiments supports 340 events.[6] We evaluated all of them and determined that resource stalls are a particularly effective DMA side effect. HPC events are more precise than TSC measurements for certain events. We assume the number of resource stalls are a direct result of the delays we can measure with TSC. As an example we present the result of a hardware performance counter called RAT_STALLS:ROB_READ_PORT (see Sect. 2.3) in Fig. 4.10. Compared to the baseline the overhead is more than double. Without our prototype malware the mean of our measurements was 1,359,898 counted events. With our prototype malware in search mode the mean was 3,161,868 counted events, and in monitor mode it was 1,535,054 counted events. The latter is only slightly higher compared to the baseline. The refined measurements demonstrate the more accurate we measure the better is the visibility of the DMA side effect.

Detection Based on our findings, DMA side effects can be measured. This means we can design a DMA malware detection mechanism. The mechanism works by establishing a measurement baseline and reference values for the TSC/HPC. During runtime, our system monitors the TSC/HPC values and compares them to the reference values. If the values deviate from the reference values DMA malware is detected. We acknowledge that an actual implementation of this delay-based detection approach needs some additional investigation. In Chap. 5 we present a more enhanced detector that is also based on HPC. Furthermore, the artificial memory stress is not required anymore to detect DMA malware with our enhanced method. In this section we discuss the I/OMMU and a detection approach based on DMA side effects as countermeasures.

4.6 Chapter Summary

In this chapter we studied DMA malware, i.e., malware hidden in dedicated hardware. Such malware can circumvent protection mechanisms run on the host CPU by directly accessing the host memory. We implemented and evaluated DAGGER, a DmA-based keystroke loGGER. The dedicated hardware enables our prototype to benefit from rootkit properties. DAGGER operates stealthily. It is undetectable by anti-virus software etc. We can conclude that DAGGER is a representative malware proof of concept when comparing it with other known DMA malware. Hence, we will reuse DAGGER in the next chapters to develop a reliable DMA malware detector.

DMA malware is more than controlling a DMA engine. Our evaluation confirmed that DMA malware is efficient even if obstacles such as memory address randomization are in place. We also demonstrated that DMA malware can be effective, that is, it can attack several OSes. This confirms that DMA malware is stealthy at no costs regarding efficiency and effectiveness. The host has no reliable means to protect itself.

---

[6] We used the Performance API, that is available at http://icl.cs.utk.edu/papi/software/index.html [accessed 25 February 2014], to work with HPC in the described experiment.

Throughout this chapter we highlighted that the I/OMMU has several issues and the host cannot necessarily count on this preventive countermeasure against DMA malware. Besides possible vulnerabilities and various preconditions that must be fulfilled for a successful I/OMMU deployment, the most obvious issue is that common OSes do not or do insufficiently support the I/OMMU. Hence, DMA malware can attack OSes such as Windows. A general and reliable approach for scanning the dedicated devices for malware does not exist. A reliable and more general DMA malware detection mechanism is needed. Other researchers have also investigated I/OMMU alternatives.

In this chapter we discussed an alternative approach. Our detection approach is based on the observation that parallel memory accesses from the isolated hardware (via DMA) and the main CPU produce measurable side effects. Hence, we can conclude that illegitimate DMA operations are not stealthy anymore. Nonetheless, we have to admit that the experimental setup used for the detection is rather artificial. We conclude that the current setup is insufficient for a detection tool that can be applied in practice. However, we demonstrated that hardware performance counters can be the basis for a reliable detection tool. We revealed that the measurement tool requires a sufficient measurement resolution. Hardware performance counters fulfill this requirement. We will further investigate this point in more detail in the next chapter.

Without an alternative, only dedicated hardware whose inner workings is accessible by the host, i.e., complete RAM and ROM access, should be deployed. This enables the host to check the device for malicious modifications from time to time. A precondition for this is a reasonable measurement strategy and that the scanner gets loaded first. Devices with a dedicated processor, dedicated runtime memory, and a DMA engine are a threat for the host platform. This chapter demonstrates that additional protection mechanisms are needed to ensure a platform's confidentiality, integrity, and especially its trustworthiness.

Chapter 5
A Primitive for Detecting DMA Malware

> You can't defend. You can't prevent. The only thing you can do is
> to detect and respond.
>
> Bruce Schneier,
> American Cryptographer,
> Computer Security and Privacy Specialist

The previous chapters presented that computer platform peripherals, or more precisely, dedicated hardware such as network interface cards, video cards and management controller can be exploited to attack the host computer platform. The dedicated hardware provides the attacker with a separate execution environment that is not considered by state-of-the-art anti-virus software, intrusion detection systems, and other system software security features available on the market. Hence, dedicated hardware is well-suited for stealthy attacks [35, 36, 46, 123, 134, 135]. Such attacks have also been integrated into exploitation frameworks [18, 19].

For example, Duflot et al. [47] presented an attack based on a network interface card (NIC) to run a remote shell to take-over the host. They remotely infiltrated the NIC with the attack code by exploiting a security vulnerability. Triulzi [134, 135] demonstrated how to use a combination of a NIC and a video card (VC) to access the main memory that enables an attacker to steal cryptographic keys and other sensitive data. Triulzi remotely exploited the firmware update mechanism to get the attack code on the system.

In Chap. 4 we described how we exploited a micro-controller that is integrated in the computer platform's memory controller hub (MCH) to hide a keystroke code logger that captures secret data, e.g., passwords. All these attacks have in common that they have to access the main memory via direct memory access. By doing so, the attacks circumvent hardened security mechanisms that are set up by host system software. Furthermore, the attack does not need to exploit a host system software vulnerability. Devices that are capable of executing DMA transactions are called bus masters, see Sect. 2.5. The host CPU that usually executes security software to reveal attacks, does not necessarily have to be involved when other bus masters access the main memory, see Chap. 4. Due to modern bus architectures, such as peripheral component interconnect express (PCIe), a sole central DMA controller, which must be configured by the host CPU, became obsolete. Firmware executed

© Springer International Publishing Switzerland 2015
P. Stewin, *Detecting Peripheral-based Attacks on the Host Memory*,
T-Labs Series in Telecommunication Services, DOI 10.1007/978-3-319-13515-1_5

in the separate execution environment of the dedicated hardware can configure the peripheral's DMA engine to read from or to write to arbitrary main memory locations. This is *invisible* to the host CPU.

In this chapter we present our *Bus Agent Runtime Monitor* (BARM)—a monitor that reveals and halts stealthy peripheral-based attacks on the platform's main memory. We developed BARM to demonstrate that the host CPU is able to detect additional (malicious) accesses to the platform's main memory that originate from platform peripherals, even if the host CPU is unable to access the isolated execution environment of the suspicious peripheral. With additional access we mean access that is not intended to deliver data to or to transfer data on behalf of the host system software. BARM is based on a primitive that is able to analyze memory bus activity. It compares actual bus activity with bus activity that is expected by host system software such as the operating system or the hypervisor. BARM reports an attack based on DMA if it detects more bus activity than expected by the host system software. BARM is also able to identify the malicious peripheral.

In the previous chapters we also presented that several preventive approaches concerning DMA attacks have been proposed. For example, Intel developed an input/output memory management unit (I/OMMU) and calls the technology Intel virtualization technology for directed I/O (VT-d [2]). The I/OMMU can be applied to restrict access to the main memory. The aim of VT-d is to provide hardware supported virtualization for the popular x86 platform. Unfortunately, I/OMMUs cannot necessarily be trusted as a countermeasure against DMA attacks for several reasons. For instance, the I/OMMU (i) must be configured flawlessly [83], (ii) can be successfully attacked [111, 146–148], and (iii) cannot be applied in case of memory access policy conflicts, see Chap. 4. Furthermore, I/OMMUs are not supported by every chipset and system software (e.g., Windows Vista and Windows 7). Another preventive approach is to check the peripheral firmware integrity at load time. Unfortunately, such load time checks do not prevent runtime attacks. Repeating the checks permanently to prevent runtime attacks is borne at the cost of system performance. Note, this also does not necessarily capture transient attacks. Furthermore, it is unclear if the host CPU has access to the whole read-only memory that stores the peripheral's firmware.

We address the challenge of detecting malicious DMA with a primitive that runs on the host CPU in this chapter. By monitoring bus activity our method does not require to access the peripheral's ROM or its execution environment. Our primitive is implemented as part of the platform's system software. The basic idea is: The attacker cannot avoid causing additional bus activity when accessing the platform's main memory. This additional bus activity is the Achilles' heel of DMA-based attacks that we exploit to reveal and halt the attack. Our proof of concept implementation BARM implements a monitoring strategy that considers transient attacks. The main goal of our technique is to monitor memory access of devices connected to the memory bus. Especially, host CPU cores fetch data as well as instructions of a significant amount of processes. This is aggravated by the in- and output (I/O) of peripherals such as network interface cards and harddisks. BARM demonstrates how to meet these challenges.

In this chapter we present a method to detect and mitigate DMA-based attacks. Our main contributions are:

- **Model of expected bus activity and measurement of actual bus activity to reveal attacks**: A new mechanism for monitoring the complete memory bus activity via a primitive executed on the host CPU is presented in this chapter. Our method is based on modeling the expected memory bus activity. Furthermore, we present a technique for monitoring the actual bus activity. We reveal malicious memory access by calculating the difference between the modeled expected activity and the measured activity. Any additional DMA activity can be assumed to be an attack.
- **Disempowerment of malicious peripheral**: We can identify the offending peripheral. We implemented and evaluated our detection model in a PoC that we call BARM. BARM is efficient and effective enough that it can not only detect, but also eliminate DMA-based attacks before the attacker caused any damage.
- **Runtime monitor measurement strategy**: We implemented a measurement strategy for permanent runtime monitoring that considers transient attacks with negligible performance overhead due to commonly available CPU features of the x86 platform.

Finally, our solution does not require hardware or firmware modifications.

5.1 General Detection Model

Two core points are the basis for our detection model. First, the memory bus is a shared resource (see Fig. 5.1). Second, the system software, i.e., the OS, records all I/O activity in the form of I/O statistics. Bus masters (CPU and peripherals) are connected to the main memory via the memory bus. That bus provides exactly one interface to the main memory that must be shared by all bus masters, see Fig. 5.1. We see this shared resource as a kind of *hook* or as the Achilles' heel of the attacker. The

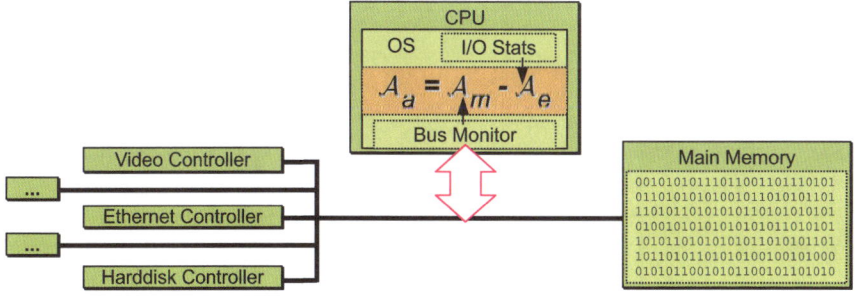

Fig. 5.1 Bus master topology exploited to reveal malicious memory access. If the difference of the measured bus activity value A_m and the expected bus activity value A_e is greater than 0, additional bus activity A_a is measured and a DMA attack is revealed

fact of the shared resource can be exploited by the host CPU to determine if another
bus master is using the bus. For example, if the host CPU cannot access the bus for a
certain amount of time, the OS can conclude that another bus master is using the bus.

How exactly the host CPU/OS determines malicious bus activity is dependent of
the implementation. We investigated multiple directions based on timing measurements and bus transactions monitoring. Experiments with the timing measurements
of bus transactions are described by Li et al. [83], for example. Timing measurements
of memory transactions are given in Sect. 4.5.2. Our experiments revealed that counting bus transaction events is the most reliable method. We present the implementation
of that novel method in Sect. 5.2.

5.2 An Implementation of the Detection Model

In this section we describe our implementation of the general detection model based
on bus transaction event counting. The purpose of our PoC implementation is to
confirm that the host CPU can detect DMA-based attacks that originate from peripherals. We implemented BARM for the Intel x86 platform. We developed BARM as
a Linux kernel module. According to the experiment described in Chap. 4, malware,
which is executed in peripherals with a separate DMA engine, can access the main
memory stealthily. The host CPU does not necessarily have to be involved when a
DMA-based memory transaction is set up. Nonetheless, the memory bus is inevitable
a shared resource that is arbitrated by the MCH, see Fig. 2.5. This is the reason why
we expect side effects when bus masters access the main memory.

We analyzed the capabilities of performance monitoring units (PMU, see Sect. 2.3)
to find and exploit such DMA side effects. PMUs are implemented as model-specific
registers. These registers can be configured to count performance related events. The
PMUs are not intended to detect malicious behavior on a computer system. Their purpose is to detect performance bottlenecks to enable a software developer to improve
the performance of the affected software accordingly [104]. In this work we exploit
PMUs to reveal stealthy peripheral-based attacks on the platform's main memory.
Malware executed in peripherals has no access to processor registers and therefore
cannot hide its activity from the host CPU by modifying the PMU processor registers. Our analysis revealed memory transaction events that can be counted by PMUs.
In particular, a counter event called BUS_TRANS_MEM summarizes all burst (full
cache line), partial read/write (non-burst) as well as invalidate memory transactions
[71]. This is the basis for BARM.

Depending on the precise processor architecture, Intel processors provide five to
seven performance counter registers per processor core [69, Sect. 18]. In this case,
at most five to seven events can be counted in parallel with one processor core.
Three of those counters are fixed function counters, i.e., the counted event cannot be changed. The other counters are general purpose counters that we use for
BARM to count certain BUS_TRANS_MEM events. We are able to successfully measure $\mathcal{A}_m$ when we apply the BUS_TRANS_MEM counters correctly. At this point, that

knowledge is insufficient to decide if the transactions exclusively relate to an OS task or if malicious transactions are also among them. In the following, we lay the groundwork to reveal malicious transactions originating from a compromised DMA-capable peripheral.

5.2.1 Bus Master Analysis

In the following we analyze the host CPU (related to the processor bus system) and the UHCI controller (related to the PCIe bus system) bus masters regarding the number of bus transactions that they cause. By doing so, we consider the most important bus systems that share the memory bus. Other bus masters, such as harddisk and ethernet controllers, can be analyzed in a similar way.

Host CPU The host CPU is maybe the most challenging bus master. The CPU causes a huge amount of memory transactions. Several processor cores fetch instructions and data for many processes. Monitoring all those processes efficiently regarding the bus activity that they cause is nearly impossible. Hence, we decided to analyze the host CPU bus agent behavior using the BUS_TRANS_MEM events in conjunction with certain control options and so-called event name extensions. We implemented a Linux kernel module for this analysis. Our key results are: (i) Bus events caused by user space and kernel space processes can be counted with one counter. (ii) The event name extensions THIS_AGENT and ALL_AGENTS can be used in conjunction with BUS_TRANS_MEM events [see 71]) to distinguish between bus transactions caused by the host CPU and all other processor bus system bus masters. THIS_AGENT counts all events related to all processor cores belonging to a CPU bus agent. ALL_AGENTS counts events of all bus agents connected to the bus where the host CPU is connected to. The ALL_AGENTS extension is very important for our implementation. It enables us to measure the bus activity value $\mathcal{A}_m$ (see Sect. 5.1) in terms of number of bus transactions:

$$\mathcal{A}_m = BUS\_TRANS\_MEM.ALL\_AGENTS \qquad (5.1)$$

Furthermore, our analysis revealed that a host CPU is not necessarily exactly one bus agent. A multi-core processor can consist of several bus agents. For example, we used a quad-core processor (Intel Core 2 Quad CPU Q9650@3.00GHz) that consists of two bus agents. Two processor cores embody one bus agent as depicted in Fig. 5.2. Hence, the number of processor cores is important when determining (il)legitimate bus transactions. Note, if the host CPU consists of several bus agents, it is necessary to start one counter per bus agent with the THIS_AGENT event name extension. With this knowledge we can determine bus master transactions of all bus masters $\mathcal{A}_m$. We can distinguish between bus activity of the host CPU (see Eq. 5.2) and bus activity caused by all other bus masters (see Eq. 5.3) that access the main memory via the MCH.

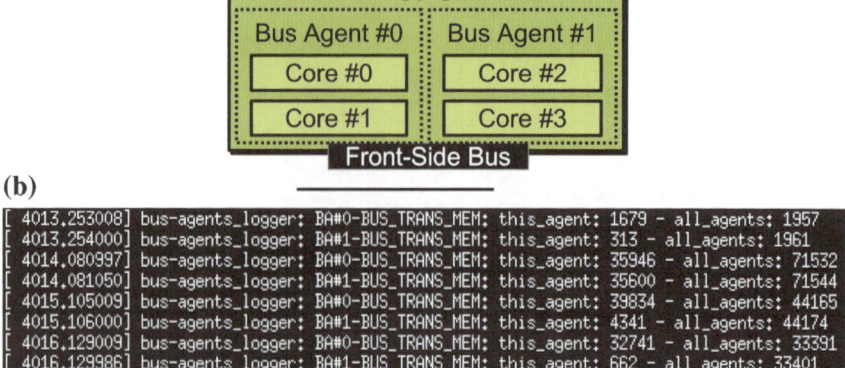

Fig. 5.2 Intel quad-core processor. The quad-core processor consists of two bus agents and each bus agent consists of two cores, see (**a**). When counting BUS_TRANS_MEM events with both bus agents, i.e., in (**b**) BA#0 and BA#1, the THIS_AGENT name extension delivers significant difference. The kernel log in (**b**) also depicts that the values for the ALL_AGENTS name extension are pretty much the same within a counter query iteration

$$\mathcal{A}_m^{CPU} = \sum_{n=0}^{H} BUS\_TRANS\_MEM.THIS\_AGENT_{cpu\_bus\_agent\#n},$$

$$H \in \mathbb{N}, H = \text{number of host CPU bus agents} - 1 \tag{5.2}$$

$$\overline{\mathcal{A}_m^{CPU}} = \mathcal{A}_m - \mathcal{A}_m^{CPU}$$

$$\Leftrightarrow \quad \mathcal{A}_m = \mathcal{A}_m^{CPU} + \overline{\mathcal{A}_m^{CPU}} \tag{5.3}$$

This means that we can subtract all legitimate bus transactions caused by user space and kernel space processes of all processor cores. Note, according to our trust and adversary model (see Sect. 2.7) the measured host CPU bus activity value and the expected host CPU bus activity value are the same ($\mathcal{A}_e^{CPU} = \mathcal{A}_m^{CPU}$), since all processes running on the host CPU are trusted. Analogously the expected bus activity value is split, i.e., $\mathcal{A}_e = \mathcal{A}_e^{CPU} + \overline{\mathcal{A}_e^{CPU}}$.

Universal Host Controller Interface Controller The *Universal Host Controller Interface* (UHCI) controller is an I/O controller for *Universal Serial Bus* (USB) devices such as a USB keyboard or a USB mouse. USB devices are polled by the I/O controller to check if new data is available. System software needs to prepare a schedule for the UHCI controller. This schedule determines how a connected USB device is polled by the I/O controller. The UHCI controller permanently checks its schedule in the main memory. Obviously, this procedure causes a lot of bus activity. Further bus activity is generated by USB devices if a poll reports that new data is available. In the following we analyze how much activity is generated, i.e., how many bytes are transfered by the UHCI controller when servicing a USB device.

In our case, the I/O controller analyzes its schedule every millisecond. That means, the controller looks for data structures that are called transfer descriptors. These descriptors determine how to poll the USB device. To get the descriptors the controller reads a frame pointer from a list every millisecond. A frame pointer (physical address) references to the transfer descriptors of the current timeframe. Transfer descriptors are organized in queues. A queue starts with a queue head that can contain a pointer to the first transfer descriptor as well as a pointer to the next queue head [see 62, p. 6]. According to Intel [62] the frame (pointer) list consists of 1024 entries and has a size of 4096 bytes. The UHCI controller needs 1,024 ms (1 entry per millisecond) for one frame (pointer) list iteration. We analyzed the number of bus transactions for one iteration with the help of the highest debug mode of the UHCI host controller device driver for Linux. In that mode schedule information are mapped into the debug file system. We determined that the frame pointers reference to interrupt transfer queues (see Fig. 5.3d: int2, int4, ..., int128) and to a queue called async. int2 means, that this queue is referenced by every second frame pointer, int4 by every fourth, int8 by every eighth, etc. The async queue is referenced by every 128th frame pointer.

Unassigned interrupt transfer queues, i.e., queues not used to poll a USB device, are redirected to the queue head of the async queue, see Fig. 5.3b. Parsing the async queue requires three memory read accesses as illustrated in Fig. 5.3a. Parsing interrupt transfer queues that are assigned to poll a USB device needs more than four memory reads. The exact number of memory reads depends on how many elements the queue has. Usually, it has one element if the queue is assigned to a USB keyboard. The queue can also have two elements if the queue is assigned to a keyboard and mouse, for example. If the queue has one element, parsing the whole assigned interrupt transfer queue needs six memory reads, see Fig. 5.3c. We summarize our examination as follows:

$$
\begin{aligned}
\#bus\ read\ transactions = {} & 8 \times \#async\ reads + 8 \times \#int128\ reads \\
& + 16 \times \#int64\ reads + 32 \times \#int32\ reads + 64 \times \#int16\ reads \quad (5.4) \\
& + 128 \times \#int8\ reads + 256 \times \#int4\ reads + 512 \times \#int2\ reads
\end{aligned}
$$

In total 4216 bus read transactions are calculated if int16 is assigned to a USB keyboard, as depicted in Fig. 5.3d. According to Intel [62], the UHCI controller updates queue elements. We expect this for the queue element of the int16 queue. This queue is referenced by 64 frame pointers. Hence, we calculate with 64 memory write transactions. This means that the overall number of bus transactions is 4280. We successfully verified this behavior with a Dell USB keyboard as well as a Logitech USB keyboard in conjunction with the single step debugging mode of the UHCI controller [see 62, p. 11], the information was retrieved from the Linux debug file system in /sys/kernel/debug/usb/uhci/, and performance monitoring units counting BUS_TRANS_MEM events.

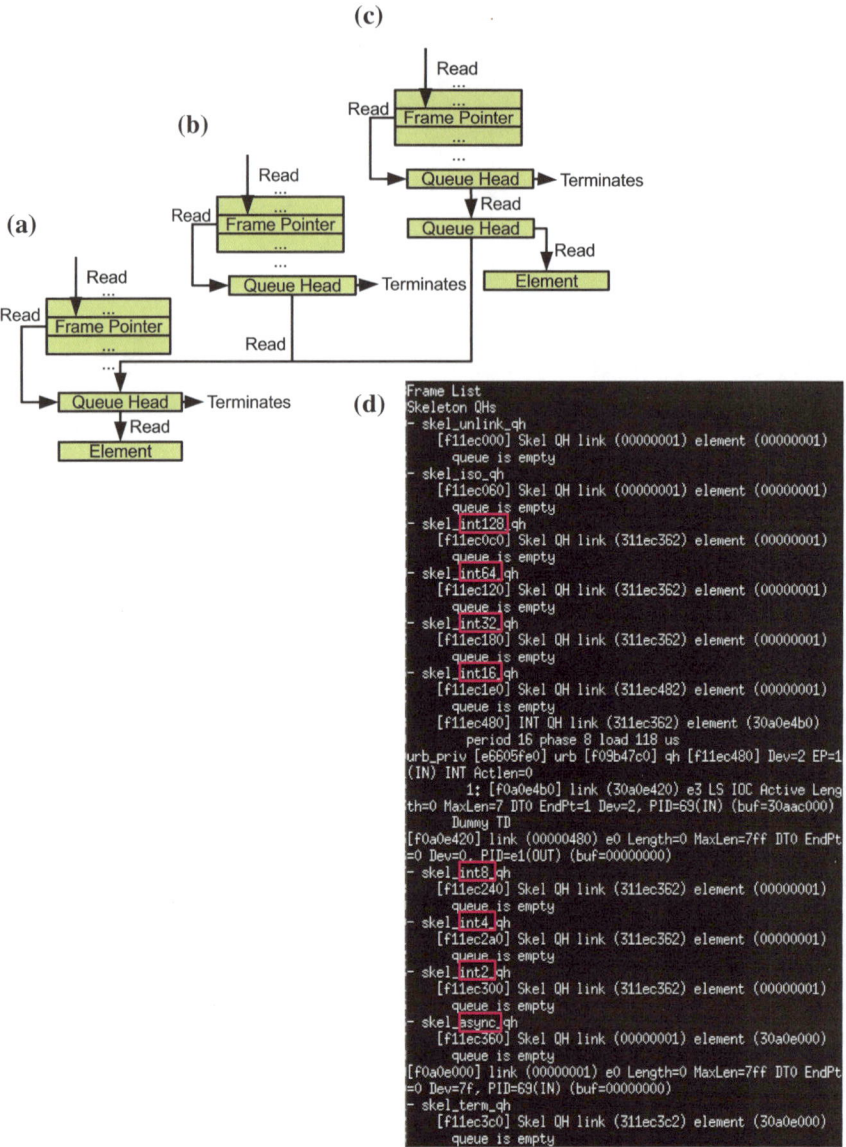

Fig. 5.3 UHCI schedule information (simplified). The schedule reveals that `int` and `async` queues are in use. The physical addresses of queue link targets are denoted in brackets. A queue link or queue element, which terminates, contains the value `00000001` instead of a physical address. The `int16` queue is responsible for our USB keyboard

With the same setup we determined how many bus transactions are needed when the USB device has new data that are to be transmitted into the main memory. For our USB keyboard we determined that exactly two bus transactions are needed to handle a keypress event. The same is true for a key release event. The Linux driver handles such events with an interrupt routine. Hence, to determine the expected bus activity $\mathcal{A}_e^{UHCI}$ we request the number of handled interrupts from the OS and duplicate it. This means that the overall number of bus transactions in our example is $\mathcal{A}_e^{UHCI} = 4280 + 2 \times \#USB$ *interrupts*.

Additional Bus Masters To handle the bus activity of the whole computer platform, the behavior of all other bus masters, such as the ethernet controller and the harddisk controller, must also be analyzed similar to the UHCI controller. We had to analyze one more bus master when we tested our detection model on Lenovo Thinkpad laptops. We were unable to turn off the fingerprint reader (FR) via the BIOS on an older Thinkpad model. Hence, we analyzed the fingerprint reader and considered this bus master for our implementation. We determined that it causes 4 bus transactions per millisecond. For this work, or more precisely, to demonstrate that the host CPU can detect DMA attacks, it is sufficient to consider up to five bus masters for BARM. Besides from the two CPU-based bus masters and the UHCI controller we also consider Intel's Manageability Engine (ME) as a bus master. During normal operation we assume $\mathcal{A}_e^{ME} = 0$. To be able to demonstrate that our detection model works with a computer platform we do not use all bus masters available on the platform in our experiment. For example, we operate the Linux OS from the computer's main memory in certain tests of our evaluation (see Sect. 5.3). This allows us to make use of the harddisk controller I/O functionality as needed.

With the analysis presented in this section we can already determine which bus master caused what amount of memory transactions. This intermediate result is depicted in Fig. 5.4.

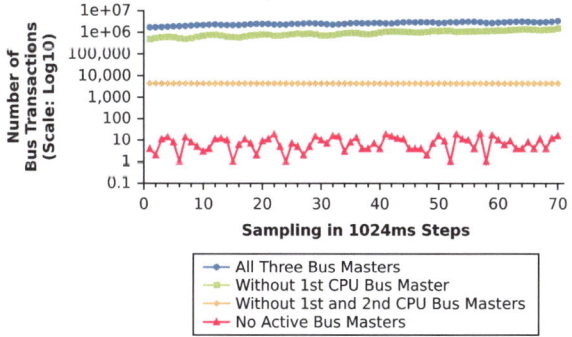

Fig. 5.4 Breakdown of memory transactions caused by three active bus masters. The curve at the *top* depicts the number of all memory transactions of all active bus masters (in our setup), that is, $\mathcal{A}_m$. The curve *below* depicts $\mathcal{A}_m$ reduced by the expected memory transactions of the first CPU bus master, that is, $\mathcal{A}_m - \mathcal{A}_e^{CPU\,BA\#0}$. The next curve *below* represents $\mathcal{A}_m - \mathcal{A}_e^{CPU\,BA\#0} - \mathcal{A}_e^{CPU\,BA\#1}$. The curve at the *bottom* represents $\mathcal{A}_m - \mathcal{A}_e^{CPU\,BA\#0} - \mathcal{A}_e^{CPU\,BA\#1} - \mathcal{A}_e^{UHCI}$

5.2.2 Bus Agent Runtime Monitor

With the bus master analysis that we introduced in Sect. 5.2.1 we were able to implement BARM in the form of a Linux kernel module. In this section we describe how we implemented a monitoring strategy that permanently monitors and also evaluates bus activity. The performance monitoring units are already configured to measure BUS_TRANS_MEM events. The permanent monitoring of $\mathcal{A}_m$, i.e., $\mathcal{A}_m^{CPU}$ and $\overline{\mathcal{A}_m^{CPU}}$, is implemented using the following steps:

1. Reset counters and store initial I/O statistics of all non-CPU bus masters (e.g., UHCI, FR, ME, HD, ETH, VC).
2. Start counting for a certain amount of time t (implemented using high precision timer).
3. Stop counters when time t is reached.
4. Store counter values for $\mathcal{A}_m$ and $\mathcal{A}_m^{CPU}$ (see Sect. 5.2.1) as well as updated I/O statistics of all non-CPU bus agents.
5. Continue with step (1) and determine $\mathcal{A}_e$ in parallel by waking up the according evaluation kernel thread.

We also need to compare the measured bus activity and the expected bus activity. BARM compares $\overline{\mathcal{A}_m^{CPU}}$ and $\overline{\mathcal{A}_e^{CPU}}$ when executing the evaluation kernel thread as follows:

1. Determine $\overline{\mathcal{A}_m^{CPU}}$ using the stored counter values for $\mathcal{A}_m$ and $\mathcal{A}_m^{CPU}$ (see Sect. 5.2.1).
2. Calculate $\overline{\mathcal{A}_e^{CPU}}$ with $\mathcal{A}_e^{UHCI}$, $\mathcal{A}_e^{FR}$, $\mathcal{A}_e^{ME}$, $\mathcal{A}_e^{HD}$, $\mathcal{A}_e^{ETH}$, $\mathcal{A}_e^{VC}$, etc., which are derived from the difference of the stored updated I/O statistics and the stored initial I/O statistics. Note, for our implementation we assume $\mathcal{A}_e^{HD} = 0$, $\mathcal{A}_e^{ETH} = 0$, etc.
3. Compare $\overline{\mathcal{A}_m^{CPU}}$ and $\overline{\mathcal{A}_e^{CPU}}$, report results and, if necessary, apply a defense mechanism.

Tolerance Value   For practicality we need to redefine how $\mathcal{A}_a$ is calculated. We use $\mathcal{A}_a$ to interpret the PMU measurements in our PoC implementation. One reason is that PMU counters cannot be started/stopped simultaneously. Very few processor cycles are needed to start/stop a counter and counters are started/stopped one after another. The same can occur in the very short amount of time, where the counters are stopped to be read and to be reset (see timeframe between step (3) and step (2) when permanently monitoring). Similar inaccuracies can occur when reading OS I/O statistics. Hence, we introduce the tolerance value $\mathcal{T} \in \mathbb{N}$ and refine $\mathcal{A}_a$:

$$\mathcal{A}_{\mathcal{T}a} = \begin{cases} 0, & \text{if } |\mathcal{A}_m - \mathcal{A}_e| \in \{0, \dots, \mathcal{T}\} \\ |\mathcal{A}_m - \mathcal{A}_e|, & \text{if } |\mathcal{A}_m - \mathcal{A}_e| \notin \{0, \dots, \mathcal{T}\} \end{cases} \tag{5.5}$$

The value of $\mathcal{T}$ is a freely selectable number in terms of bus transactions that BARM can tolerate when checking for additional bus traffic. Our evaluation

Best Case for Attacker

Fig. 5.5 Tolerance value $\mathcal{T}$. If the attacker can predict the very exact moment where BARM determines $\mathcal{T}$ too little bus transactions, an attack with $2\mathcal{T}$ bus transactions could theoretically executed stealthily

demonstrates that a useful $\mathcal{T}$ is rather a small value (see Sect. 5.3). Nonetheless, we have to consider that $\mathcal{T} > 0$ theoretically gives the attacker the chance to hide the attack, i.e., to execute a transient attack. In the best case (see Fig. 5.5) the stealthy attack can have $2\mathcal{T}$ bus transactions at most. It is very unlikely that $2\mathcal{T}$ bus transactions are enough for a successful attack. Data is most likely at a different memory location after a platform reboot. Hence, the memory must be scanned for valuable data and this requires a lot of bus transactions. Mechanisms such as address space layout randomization (ASLR, see also Sect. 4.3.3) that are applied by modern OSes can also complicate the search phase. This results in additional bus transactions. Furthermore, the attacker needs to know the very exact point in time when BARM must tolerate $-\mathcal{T}$ transactions.

Identifying and Disabling the Malicious Peripheral If $\mathcal{A}_{\mathcal{T}a} > 0$ BARM has detected a DMA-based attack originating from a platform peripheral. It is already of great value to know that such an attack is executed. A simple defense policy that can be applied to stop an attack is to remove bus master capabilities using the BME bit (see Sect. 2.5) of all non-trusted bus masters. Such a policy can be insufficient if all platform features are required for operation. It could also result in data loss without further measures. However, a system that has been compromised via such a targeted attack should be taken offline for a detailed examination.

When stopping the non-trusted bus masters BARM places a notification for the user on the platform's screen. $\mathcal{A}_{\mathcal{T}a}$ does not include any information about what platform peripheral is performing the attack. To include that information in the notification message, we implemented a simple peripheral test that identifies the suspicious peripheral. When the DMA attack is still scanning for valuable data, we unset the BME bits of the non-trusted bus masters one after another to reveal the malicious peripheral. After the bit is unset, BARM checks if the additional bus activity vanished. If so, the malicious peripheral is identified and the peripheral name is added to the attack notification message. If BARM still detects additional bus activity the BME bit of the wrong peripheral is set again. The OS must not trigger any I/O tasks during the peripheral test. Our evaluation reveals that our test is performed in a few milliseconds, see Sect. 5.3. It is required that the attack is a bit longer active than our peripheral test. Otherwise, it cannot be guaranteed that our test identifies the malicious peripheral. The DMA attack on a Linux system described in Chap. 4 needs between 1,000 and 30,000 ms to scan the memory. Our evaluation demonstrates that BARM can detect and stop a DMA attack much faster.

5.3 Evaluation of the Detection Model Implementation

We evaluated BARM, which is implemented as a Linux kernel module. First, we conducted tests to determine a useful tolerance value T. In the main part of this section, we present the performance overhead evaluation results of our solution. We demonstrate that the overhead caused by BARM is negligible. Finally, we conducted some experiments to evaluate how BARM behaves during an attack.

5.3.1 Tolerance Value T

We performed several different tests to detemine a useful tolerance value. We repeated each test 100 times. Several different tests means, we evaluated BARM with different PMU value sampling intervals (32, 128, 512, 1,024, 2,048 ms), number of CPU cores (1—4cores), RAM size (2gigabyte, 4gigabyte, 6gigabyte, 8gigabyte), platforms (Intel Q35 Desktop / Lenovo Thinkpads: T400, X200, X61s), as well as minimum (power save) and maximum (performance) CPU frequency to check the impact for T. Furthermore, we evaluated BARM with a CPU and with a memory stress test. CPU stress test means, running the `sha1sum` command on a 100 megabyte test file 100 times in parallel to ensure that the CPU utilization is 100 %. For the memory stress test, we copied the 100MB test file 2,000 times from a main memory location to another. Our platforms had the following configurations: Q35–Intel Core 2 Quad CPU Q9650@3.00GHz with 4gigabyte RAM, T400–Intel Core 2 Duo CPU P9600@2.66GHz with 4gigabyte RAM, X200–Intel Core 2 Duo CPU P8700@2.53GHz with 4gigabyte RAM, and X61s–Intel Core 2 Duo CPU L7500@1.60GHz with 2gigabyte RAM. We used a sampling interval of 32 ms, 1core, 4gigabyte RAM, the Q35 platform, and the maximum CPU frequency as basic evaluation configuration. We only changed one of those properties per test. The results are summarized in Fig. 5.6.

Note, to determine T we considered up to five bus masters (1–2 CPU, 1 UHCI, 1 fingerprint reader, and 1 ME bus master). We used the SliTaz Linux distribution[1] that allowed us to run the Linux operating system from RAM. As a result we were able to selectively activate and deactivate different components as the harddisk controller bus master. The overall test results reveal a worst case discrepancy between measured and expected bus transactions of 19 (absolute value). This result confirms that the measurement and evaluation of bus activity yields reliable values, i.e., values without hardly any fluctuations. Nonetheless, to be on the safe side we work with a tolerance value $T = 50$ when we evaluate BARM with a stealthy DMA-based keystroke logger, see Sect. 5.3.3.

---

[1] See http://www.slitaz.org/ [accessed 25 February 2014].

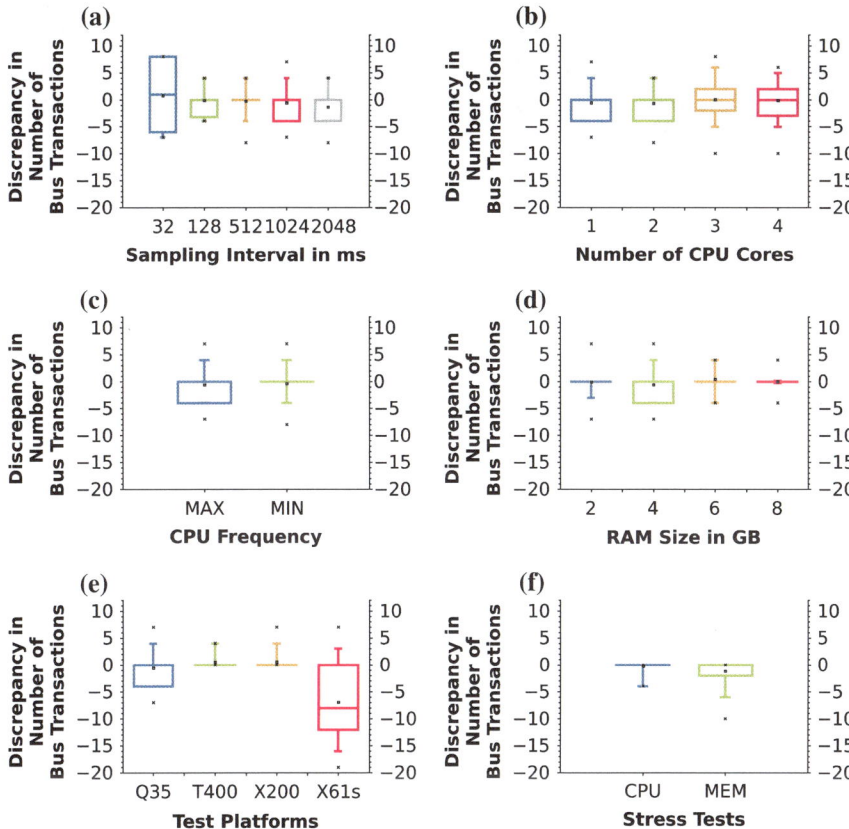

Fig. 5.6 Determining an adequate tolerance value $\mathcal{T}$. **a–f** present the discrepancy of $\mathcal{A}_a$ computations when evaluating BARM with different tests. BARM performed 100 runs on each test to determine $\mathcal{A}_a$. With discrepancy we mean the difference between the maximum and minimum $\mathcal{A}_a$ value. **a–f** visualize the discrepancy in the form of boxplots. For each test the respective minimum, lower quartile, median, upper quartile as well as maximum $\mathcal{A}_a$ value is depicted. The small point between minimum and maximum is the average $\mathcal{A}_a$ value. The $\mathcal{A}_a$ values range mostly between -10 and 10. The highest absolute value is 19, see **e** X61s

5.3.2 Performance Overhead When Permanently Monitoring

Since BARM affects only the host CPU and the main memory directly, we evaluated the performance overhead for those two resources. BARM does not access the harddisk and the network card when monitoring. We evaluated BARM on a 64bit Ubuntu kernel (version 3.5.0-26). During our tests we run the host CPU with maximum frequency thereby facilitating the host CPU to cause as much bus activity as possible. Furthermore, we executed our test with 1 CPU bus master as well as with 2 CPU bus masters to determine if the number of CPU bus masters has any impact on the performance overhead. Eventually, we need to use more processor registers

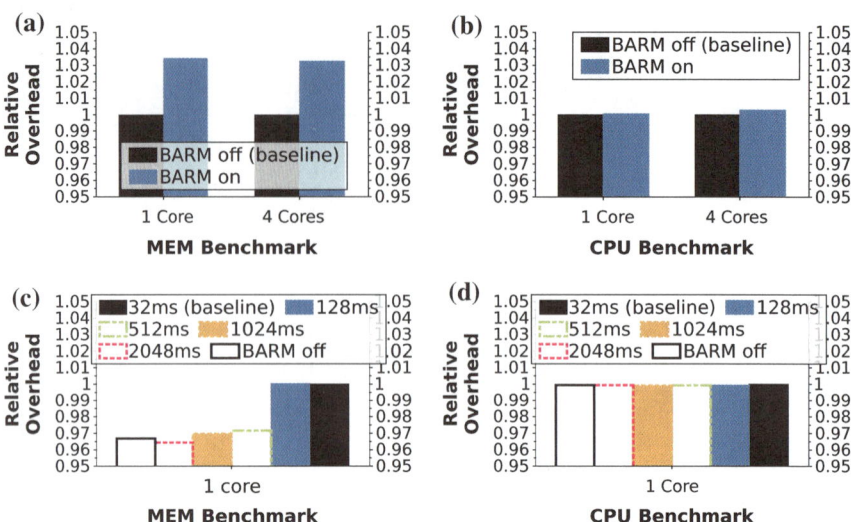

Fig. 5.7 Host performance CPU and MEM overhead evaluation. We measured the overhead with a memory (MEM) and a CPU benchmark, each passed with 1 online CPU core (1 CPU bus master) and 4 online CPU cores (2 CPU bus masters), see (**a**) and (**b**). At first, we performed the benchmarks without BARM to create a baseline. Then, we repeated the benchmarks with BARM (sampling interval: 32 ms). The results are represented as the relative overhead. The CPU benchmark did not reveal any significant overhead. The MEM benchmark revealed an overhead of approx. 3.5 %. The number of online CPU cores/CPU bus masters has no impact regarding the overhead. Furthermore, we checked the overhead when running BARM with different sampling intervals, see (**c**) and (**d**). Again, the CPU benchmark did not reveal any overhead. The MEM benchmark results reveal that the overhead can be reduced when choosing a longer sampling interval. A longer interval does not prevent BARM from detecting a DMA attack. A longer interval *can* mean that the attacker caused some damage before the attack is detected and stopped

(PMUs) with a second CPU bus master. Another important point is the evaluation of the sampling interval. Hence, we configured BARM with different intervals and checked the overhead. To measure the overhead we used time stamp counters (see Sect. 2.3) for all our tests. The evaluation results are depicted in Fig. 5.7.

5.3.3 A Use Case to Demonstrate BARM's Effectiveness

Even if we do not consider all platform bus masters in our presented PoC implementation we can demonstrate the effectiveness of BARM. This is possible because not all platform bus masters are needed for every sensitive application. For example, when the user enters a password or other sensitive data, only the UHCI controller and the CPU are required. We evaluated BARM with password prompts on Linux. We set up an environment where four bus masters are active (2 CPU, 1 UHCI, and 1 ME bus master) when using the sudo or ssh command. BARM was started together with the sudo or ssh command and stopped when the password had been entered. BARM stopped unneeded bus masters and restarted them immediately after the password prompt had been passed. We attacked the password prompt with our DMA-based keystroke logger DAGGER, which is executed on Intel's ME, see Chap. 4. DAGGER scans the main memory via DMA for the physical address of the keyboard buffer, which is also monitored via DMA.

Figure 5.8a visualizes the measurements taken by BARM when the platform is under attack. Under attack means that DAGGER is already loaded when the user is asked for the password. Figure 5.8b depicts the results of BARM when the platform is attacked at an arbitrary point during runtime. For comparison Fig. 5.8a, b also visualize BARM's measurements when the platform is not attacked. Figure 5.8c is a fraction of the kernel log, which confirms how fast BARM stopped DAG-GER. BARM detected the DMA attack at time stamp 350.40104 s. At time stamp 350.465042 s BARM identified the malicious DMA-based peripheral. This test confirms that BARM can even detect attacks before the attacker does damage. BARM stopped the attack when the keystroke logger was still in the search phase. This means that the keystroke logger did not find the keyboard buffer. Hence, the attacker was unable to capture any keystrokes. We configured BARM with a PMU value sampling interval of 32 ms. Our evaluation revealed that the attacker already generated more

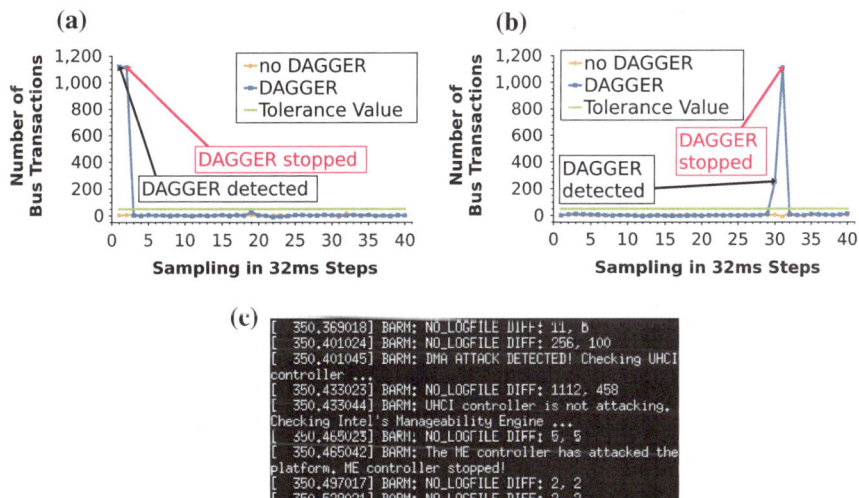

Fig. 5.8 Evaluating BARM with password prompts (ssh command) and at an arbitrary point during runtime. BARM checks for additional bus activity $\mathcal{A}_a$ every 32 ms (sampling interval). $\mathcal{A}_a$ is found if the measured value is above the tolerance value $\mathcal{T} = 50$. When the platform is not attacked the values are below $\mathcal{T}$, see (**a**) and (**b**) "no DAGGER". **a** depicts an attack where DAGGER is already waiting for the user password. BARM detects DAGGER with the first measurement and stops it almost immediately. **b** presents DAGGER's attempt to attack the platform at an arbitrary point during runtime with a similar result. **c** is the kernel log generated by BARM during the attack attempt presented in (**b**)

than 1,000 memory transactions in that time period. This means that we could have chosen even a significantly higher tolerance value than $T = 50$ bus transactions.

5.4 Limitations of Current BARM Implementation

Although the evaluation of the implementation of the detection model demonstrated that DMA malware can be found with negligible performance overhead, the current implementation also has several limitations. So far, we only considered a certain UHCI controller, a fingerprint reader, two CPU bus agents, and a ME peripheral as bus masters (see Sect. 5.3). Even though we know that each bus master can only access the main memory via one interface, we cannot exclude the possibility that the current approach for the detection model is insufficient to integrate all possible bus masters. The currently integrated bus masters are sufficient to demonstrate that BARM can detect DAGGER. Additionally, we only consider a single generation of Intel chipsets. This means that additional investigations with other chipset generations as well as chipsets made by other manufacturers are necessary to determine the extent to which BARM is generic.

Another limitation is the fact we tested BARM with one DMA malware example. Although DAGGER represents typical DMA malware, i.e., it has to search for valuable data in the host memory, does not require any cooperation with host software, and accesses the main memory via the system memory interface, we cannot exclude that other DMA malware implements mechanisms to circumvent BARM. Theoretically, an adversary could try to exploit the $2T$ bus transaction range per sampling interval (see Fig. 5.5), for example. This means that the adversary could hide up to $2T$ bus transactions if it is possible to predict the very exact moment where BARM determines T too little bus transactions.

However, if the adversary finds a way to exploit the $2T$ bus transaction range per sampling interval, this would also result in a slower search phase to find valuable host data. The amount of $2T$ bus transactions is significantly lower compared to the amount of bus transactions that are usually available to the adversary in a sampling interval. Conversely, depending on the search time for the target data in the host memory, the host CPU could exploit the delayed search to, e.g., rearrange the memory address space. This would enforce the adversary to restart the search phase. Hence, BARM should be tested with additional DMA malware examples to confirm that BARM can also detect DMA malware other than DAGGER.

Bus masters such as the ethernet controller could try to circumvent BARM by (i) ignoring the source address of the data to be copied via DMA and (ii) exploiting the number of bus transactions determined by the length of the data to be copied for attacking the host memory. The source address as well as the length are provided by the host when the host wants to send a network packet, for example. The adversary only exploits the number of bus transactions determined by the length. Hence, BARM will not detect any additional bus activity, since the adversary camouflages the illegitimate bus transactions as expected bus transactions. This kind of attack can be considered a MitM attack conducted by the network interface card. To be able to successfully conduct this MitM attack the attacker also needs to correctly determine the number of expected ethernet controller bus transactions. Chapter 6 presents how to consider MitM attacks conducted by the network interface card. The chapter also demonstrates that it is difficult to calculate the correct number of expected ethernet controller bus transactions (compared to the UHCI controller). Hence, the adversary must consider potential performance overhead caused when calculating expected ethernet controller bus transactions.

5.5 Chapter Summary

In this chapter we demonstrate that the host CPU is able to detect additional, i.e., stealthy and malicious main memory accesses that originate from compromised peripherals. The basic idea is that the memory bus is a shared resource that the attacker cannot circumvent to attack the platform's main memory. This is the attacker's Achilles' heel that we exploit for our detection method. We compare the expected bus activity, which is known by the host system software, with the actual bus activity. The actual bus activity can be monitored due to the fact that the bus is a shared resource. We developed the PoC implementation BARM and evaluated our method with up to five bus masters considering the most important bus systems (PCIe, FSB, memory bus) of a modern computer platform. BARM can also identify the compromised peripheral and disable it before the device causes any damage.

Since the host CPU can detect DMA attacks, we conclude that the host CPU can defend itself without any firmware and hardware modifications. The platform user does not have to rely on preventive mechanisms such as an I/OMMUs. We chose to implement a runtime monitoring strategy that permanently monitors bus activity. Our monitoring strategy considers transient attacks. The countermeasures presented in the related work chapter (see Sect. 3.2) such as signed firmware and latency-based attestation do not consider transient attacks. BARM can be implemented with less effort and without detailed knowledge of the inner workings of the peripheral's firmware and hardware compared to latency-based attestation approaches, see Sect. 3.2.3.

We also identified limitations of the current BARM implementation such as the theoretically exploitable bus transaction range per sampling interval ($2\mathcal{T}$) or a possible MitM attack conducted by the ethernet controller. BARM is unable to detect a MitM attack implemented in the network interface card that could be revealed with a latency-based attestation approach. Such attacks can also be prevented by applying end-to-end security in the form of a trusted channel [52]. We adapt the concept of a trusted channel in Chap. 6 to enable BARM to detect MitM attacks. Nevertheless, our BARM evaluation demonstrates that the performance overhead is negligible. Hence, we conclude that our method can be deployed in practice.

Chapter 6
Authentic Reporting to External Platforms

> *Using encryption on the Internet is the equivalent of arranging*
> *an armored car to deliver credit card information from someone*
> *living in a cardboard box to someone living on a park bench.*
> Gene Spafford,
> Professor of Computer Science

Our motivation for implementing an authentic channel application for state reporting is to deliver BARM's measurement results to an external platform protected from DMA malware. The external communication partner can evaluate the transmitted measurements to check if the counterpart has been attacked by DMA malware. The measurement results are based on processor register values (see Sect. 5.2). To exclude malware on the network interface card from modifying and forging outgoing network packets we need a secure communication channel. Such a channel not only assures confidentiality, integrity, and freshness of the transmitted data, but also authenticity of the channel endpoints. To implement such a channel we adapt the concept of a trusted channel that we presented in prior work [10, 52].

A trusted channel is a communication channel that implements secure channel properties and additionally binds communication endpoint state information to the communication session. Deploying a secure channel based on IPsec or TLS is insufficient in our case. IPsec or TLS based secure channels ensure confidentiality, integrity and freshness of the transmitted data. However, these channels are not bound to the actual communication endpoint. We implement the trusted channel based reporting application for BARM to prevent at least the following attacks. Such attacks could be conducted by malware that is executed on the network interface card. The malware could prevent BARM from communicating with the external platform by blocking or corrupting outgoing network packets. An attacker could also use such malware to steal key material, which is present in the host main memory, of the secure channel via DMA. Afterwards, the attacker can conduct a MitM attack. The malware could also relay the platform state information of a third platform, which is not attacked by DMA malware, to the external administrator platform. This means that the administrator platform could be tricked by conducting a *relay attack*.

© Springer International Publishing Switzerland 2015   71
P. Stewin, *Detecting Peripheral-based Attacks on the Host Memory*,
T-Labs Series in Telecommunication Services, DOI 10.1007/978-3-319-13515-1_6

We require at least secure channel properties (requirement R1) to ensure confidentiality, integrity, and freshness of the transfered data for our authentic reporting channel (see [52, p. 32]). The confidentiality property ensures that the attacker only gets a minimal amount of information. The integrity property ensures that corrupted network packets will be revealed immediately. The freshness property prevents the attacker from conducting a replay attack where a valid communication session is recorded to be replayed at some later time. To reveal an attack that is blocking packets that contain platform state information we introduce so-called *heartbeat* messages as payload that has to be sent during the communication session. A heartbeat in computing is a signal that indicates that, e. g., the corresponding software is still up and running [132].

The heartbeat message consists of the current BARM measurement and log information if an attack was prevented. If the network interface card has been stopped due to an attack heartbeat messages will no longer be received by the external platform. This behavior is interpreted by the external platform as a NIC-based attack. The transmitted information also includes state changes. State changes were also considered by the trusted channel concept [10, 52], but efficient and effective runtime monitoring with negligible performance overhead as implemented in BARM was missing (see [52, p. 36]): "A state change on one platform is noticed by CM (an efficient monitoring agent assumed [...]". BARM represents the missing "monitoring agent" in our DMA malware scenario.

Compared to prior work [10, 52] the trust and adversary model for our DMA malware scenario does not require trusted computed mechanisms as proposed by the TCG, see Sect. 2.7. Our channel is not based on a TPM since we do not rely on load-time code integrity checks, see Sect. 3.2.1. Channel linkage to load-time measurements stored in a TPM is not required in our application. We require that the results determined by BARM are bound to our channel (requirement R2). This is necessary during the negotiation of the communication session as well as during the communication session itself.

Please note that we do not count on the I/OMMU such as Intel's VT-d implementation. This is another difference to the trust model of our prior work [52]. This technology was introduced shortly before our results were published [52]. This means that previous authors had not been confronted with I/OMMU issues as presented in Sect. 4.5.1. Previous works assumed that drivers capable of configuring the I/OMMU correctly exist. For this work we analyzed the I/OMMU in more detail and we decided not to rely on VT-d for our authentic reporting channel. Our prior work also introduced the requirement for privacy (requirement R3). This means, the channel considered the least information paradigm to minimize the disclosure of platform state information to only the bare necessities.

The main contributions of this chapter are as follows:

- **Authentic reporting channel that excludes the network interface card from the endpoint**: Malware executed on the network interface card is able to steal secret key material from the main memory to conduct a MitM attack. Hence, we developed an authentic reporting channel that ensures that only the host CPU is the communication endpoint. Our channel is based on the secure channel protocol TLS. We adapt the TLS protocol to exchange BARM measurements and to bind the channel to its supposed endpoint. An additional feature of our communication channel is platform state change reporting. This means that our runtime monitor BARM permanently delivers every state change regarding DMA malware to the communication partner via the authentic reporting channel. Our TLS modifications are based on TLS extensions. This means that our channel is compliant with the TLS specification. Our TLS compliant channel is the first channel that considers platform state reporting regarding DMA malware. It is also the first channel that is based on an implemented effective and efficient runtime monitor to report state changes. Previous work only assumed the presence of such a runtime monitor.
- **Analysis of the ethernet controller**: Our communication channel requires the network interface card. Hence, the ethernet controller will induce bus transactions. These bus transactions must be considered by BARM. This chapter demonstrates how the ethernet controller can be integrated into BARM's detection model, i.e., how to utilize the ethernet controller as an additional bus agent.
- **Enhancing BARM's detection model with a new parameter**: The ethernet controller transfers data packets, which size is greater than the size of address pointers and keystroke codes. We demonstrate that the cache line size is an important parameter for BARM's detection model. The cache line size is necessary to compute the number of expected bus transactions correctly.
- **Exploiting additional performance monitoring unit events**: We demonstrate that certain performance monitoring unit configurations can be exploited to distinguish between memory read bus transactions and memory write bus transactions. This enables us to check if the number of expected read bus transactions and expected write bus transactions that are caused by the ethernet controller are correctly determined by BARM's detection model.

The following section starts with a description of the authentic reporting channel model. Afterwards, we explain how we implemented this model.

6.1 Implementation Independent Model

Our channel model considers client C (target platform) and server S (external platform) communication. Each endpoint may request platform state information (i.e., BARM measurements) of the peer. A local security policy determines what exactly happens after the platform state information of the peer has been evaluated. Our authentic reporting channel is controlled by host CPU software. The channel can be

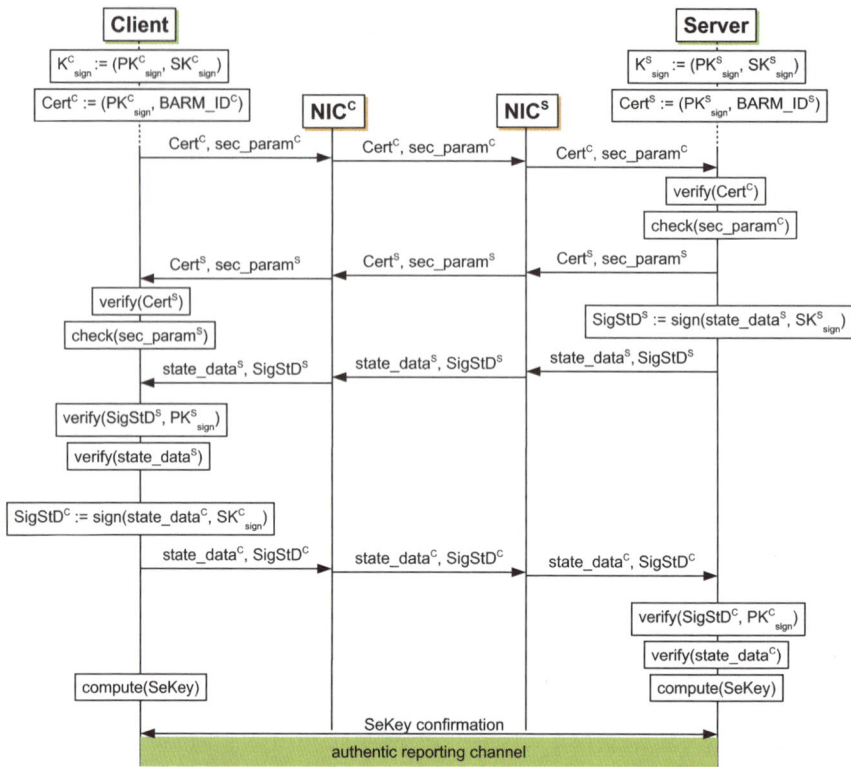

Fig. 6.1 Negotiating an authentic reporting channel. Negotiating an authentic reporting channel between Client C (target platform) and Server S (e.g., external administrator platform). *NIC*—Network Interface Card; K_{sign}—Asymmetric signing key pair (PK_{sign}, SK_{sign}) bound to; host CPU software components; *PK*—Public key; *SK*—Secret key; *Cert*—Certified public key part of key pair bound to host CPU; software components; *BARM_ID*—Host CPU software components identifier; *sec_param*—Required security parameters; *state_data*—Platform state data determined by BARM; *SigStD*—Signature of platform state data; *SeKey*—Session key

negotiated through a potentially compromised network interface card. We describe a high-level protocol for negotiating and maintaining an authentic reporting channel in the following section. Please note, in the following we omit the superscript C and S due to the symmetric protocol characteristic.

6.1.1 Negotiating an Authentic Reporting Channel

One important idea of our authentic reporting channel is to prevent platform peripherals from accessing sensitive information that is related to the channel such as secret key material. Only host CPU software is allowed to use sensitive channel information.

Please note, a peripheral could steal such information via DMA. However, BARM will reveal and stop this kind of DMA attack, see Sect. 5.3.3. Figure 6.1 depicts the handshake protocol for negotiating an authentic reporting channel for BARM. In order to conduct the handshake, both parties require a signing key K_{sign} that is an asymmetric key pair, i.e., $K_{sign} := (PK_{sign}, SK_{sign})$. Furthermore, both peers require a certificate $Cert$, which includes PK_{sign} as well as a host CPU software components identifier ($BARM\_ID$). This certificate is issued by a trusted party, which can be the external administrator platform. The signing key and the certificate are created before negotiating an authentic reporting channel. Each peer verifies the certificate including $BARM\_ID$ of its counterpart.

The creation of the channel begins with the negotiation of security parameters. This means that each party sends its certificate as well as security requirements in the form of security parameters to the peer. The security parameters determine which party reports its platform state information. Each peer checks if the security requirements of the counterpart are acceptable. In the next step, each party sends its platform state data (the current BARM measurement) to the peer. The state data is digitally signed and the corresponding signature is transmitted together with the state data. This ensures that the received state data has been sent by the expected communication partner. Both parties verify the signature with PK_{sign} that was sent by the peer as part of the certificate $Cert$. If the signature is valid both parties verify the state data. The handshake may be aborted due to DMA malware that attacks the peer. This is the case when the transmitted BARM measurement result is greater than the tolerance value $\mathcal{T}$ (see Sect. 5.2.2). After both client and server have verified the exchanged data successfully the same session key is computed and confirmed by both platforms. The computed session key will be bound to the communication session. After the confirmation the authentic communication session is in place and both peers start to periodically send heartbeat messages.

State Change The heartbeat messages either confirm the current platform state or they report a state change. The reported platform state can reveal that the peer is under a DMA malware attack, that the suspicious peripheral could be stopped, or that no attack has been detected. If the peer stops sending heartbeat messages, the local platform assumes that the peer has been attacked by DMA malware executed on the network interface card. In this case, BARM has successfully terminated the ongoing DMA attack by stopping the network interface card. Depending on the local security policy a platform can tear down the channel, continue with the current session key, or renegotiate the channel. It is advantageous to continue with the current session key if the heartbeat message reports that the attack could be stopped immediately and if the local security policy states that this case is tolerable. To be more precise, it can make sense if the platform can continue to operate normally without the affected peripheral. In the case of an involved administrator platform, we expect that the administrator will analyze the attack in more detail as soon as possible to remove the DMA malware from the compromised peripheral or, if absolutely necessary, to exchange the compromised peripheral or chipset with a benign one.

6.2 Implementation of the Authentic Reporting Channel for BARM

BARM as presented in Chap. 5 is insufficient for the authentic channel based reporting application. When BARM sends network packets, it also causes bus activity that needs to be considered by BARM's detection model. To implement an authentic channel application for our DMA malware scenario we have (i) enhanced BARM's detection model, see Sect. 6.2.1 and (ii) modified the TLS protocol to bind BARM's measurement (state information) to that channel, see Sect. 6.2.2.

6.2.1 Bus Master Analysis: Ethernet Controller

To consider the ethernet controller in BARM's detection model we have to determine the expected bus activity value $\mathcal{A}_e^{ETH}$. Hence, we conducted a similar bus master analysis as presented in Sect. 5.2.1 for the ethernet controller of our target platform. We analyzed the ethernet controller (namely *Ethernet Controller: Intel Corporation 82566DM-2 Gigabit Network Connection (rev 02)* [65]) of the same target platform as the previous experiments, see Chaps. 4 and 5. The corresponding ethernet controller Linux device driver is e1000e.ko. To simplify our analysis we configured the driver to use legacy interrupts and no interrupt delays as well as no interrupt throttling. We also disabled checksumming and segmentation offloading for the network device.

The ethernet controller works with so-called descriptor rings, i.e., the transmit descriptor ring and the receive descriptor ring, see Fig. 6.2. Each ring consists of 256 descriptors. A descriptor has a size of 16 bytes. This means that the device driver allocates 4,096 bytes for each ring. If the host intends to send network packets, it prepares transmit descriptors and informs the ethernet controller that new descriptors are ready to be processed. The ethernet controller reads the descriptors via DMA from the host memory. After evaluating the descriptor the controller copies the network packet data from the host memory address that is present in the descriptor (see Fig. 6.2) to its internal memory to be able to send the packet. If the ethernet controller has processed the descriptor, the controller "returns" the descriptor to the host by writing the descriptor done bit in the status field of the descriptor via DMA. When receiving network packets the process is similar except that the ethernet controller writes the network packet data into the host memory.

Cache Line Size To integrate the ethernet controller as a bus master into BARM's detection model we have to consider that the size of network packets is usually greater than keystroke codes, see Sect. 5.2.1. Keystroke codes are transfered via one bus transaction. This is not valid for network packets that have a size of 1,514 bytes for example. To be able to determine how many bus transactions are necessary to

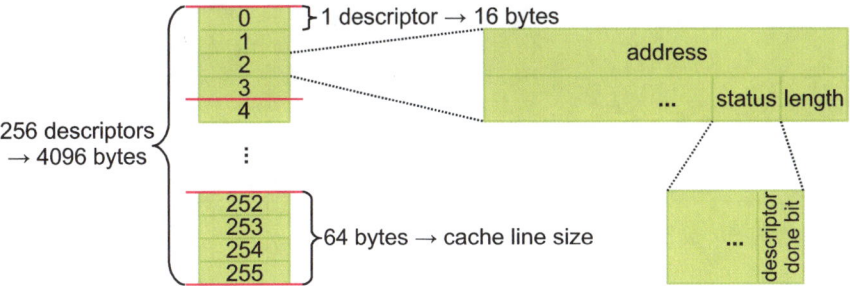

Fig. 6.2 Transmit/receive descriptor ring structure. When the device driver informs the NIC that new network packets are ready to be transmitted, the ethernet controller reads transmit descriptors from the descriptor ring. The controller also reads the corresponding packets of the size that is stored in the `length` field of the descriptor from the host memory address that is stored in the `address` field of the descriptor. The ethernet controller writes the `descriptor done bit` in the `status` field of the descriptor if the the descriptor has been processed. When new network packets arrive from the network, the ethernet controller reads receive descriptors from the descriptor ring. Afterwards, the controller writes the corresponding packets of the size that is stored in the `length` field of the descriptor to the host memory address. The address is stored in the `address` field of the descriptor. The ethernet controller writes the `descriptor done bit` in the `status` field of the descriptor if the the descriptor has been processed

transfer a particular amount of data we introduce a new parameter, i.e., the *cache line size*. The system cache is organized in cache lines. Memory accesses are handled in cache lines of a certain cache line size $C \in \mathbb{N}$ (see [127, p. 223]). C is 64 bytes for our platform (see [63, p. 17]). That means, if one word is requested from main memory, 64 bytes are actually transfered in one memory transaction. It is assumed that data that is adjacent in the host memory will likely be accessed in a subsequent operation. If so, these bytes are already in the cache and no additional transaction is needed. Memory access of peripherals is also handled in cache lines. It is possible that such a transaction must be snooped to ensure a coherent cache line (see [63, p. 27]).

The descriptor dump of the `e1000e.ko` driver depicts the host memory addresses of the network packet data, see Fig. 6.3. The dump also reveals that not every address is cache line size aligned. This means that the number of bus transactions required to transfer the network packet data via DMA is not necessarily the value stored in the `length` field divided by the cache line size. Another important point relates to the receive descriptor handling. According to Intel [65] the ethernet controller optimizes the process of returning receive descriptors. That means, when receiving packets the ethernet controller does not write the `descriptor done bit` for each descriptor individually. Instead, it "collects" four descriptors that belong to the same cache line to be able to write four `descriptor done bits` with one bus transaction, see Fig. 6.2. We consider both scenarios for the equation to compute the expected bus transactions caused by the ethernet controller.

Fig. 6.3 Transmit descriptor/receive descriptor dump of the `e1000e.ko` driver. The dump reveals the most important information to derive the number of bus transactions caused by the ethernet controller. Some host memory addresses are not cache line size aligned. This can result in an additional bus transaction

Expected Bus Activity of the Ethernet Controller Due to our analysis we define the expected bus activity of the ethernet controller as follows:

$$\mathcal{A}_e^{ETH} = \mathcal{A}_e^{TX_{reads}} + \mathcal{A}_e^{TX_{writes}} + \mathcal{A}_e^{RX_{reads}} + \mathcal{A}_e^{RX_{writes}} \tag{6.1}$$

$\mathcal{A}_e^{TX_{reads}}$ is the expected bus activity that is caused by memory reads when transmitting a packet. $\mathcal{A}_e^{TX_{writes}}$ represents activity that is caused by memory writes. Analogously, $\mathcal{A}_e^{RX_{reads}}$ and $\mathcal{A}_e^{RX_{writes}}$ are introduced to consider the bus activity when receiving network packets. To compute $\mathcal{A}_e^{TX_{reads}}$, $\mathcal{A}_e^{TX_{writes}}$, $\mathcal{A}_e^{RX_{reads}}$, and $\mathcal{A}_e^{RX_{writes}}$ for one BARM sampling interval we have to consider the cache line size for the memory buffers that are read and written. That means for the memory buffer that stores the network packet data in host memory we have to align the memory buffer start address, which is stored in the `address` field ($hma \in \mathbb{N}$) of a descriptor, to the previous cache line size aligned address. The result is $ba\_start \in \mathbb{N}$:

$$ba\_start = hma - (hma \bmod \mathcal{C}) \tag{6.2}$$

The alignment for the memory buffer end address ($ba\_end \in \mathbb{N}$), which is the sum of the value in the `address` field (hma) and the value of the `length` field ($len \in \mathbb{N}$) of a descriptor is as follows:

$$ba\_end = hma + len + \mathcal{C} - ((hma + len) \bmod \mathcal{C}) \tag{6.3}$$

The same alignment is required for descriptor transfers. The transfer start address is determined by the descriptor number of the last descriptor of the previous sampling interval ($old\_d \in \mathbb{N}$). The transfer end address is determined by the descriptor number of the last descriptor of the current sampling interval ($cur\_d \in \mathbb{N}$). When considering the cache line size the alignment results in descriptor numbers $d\_start \in \mathbb{N}$ and $d\_end \in \mathbb{N}$ as follows ($\mathcal{D} \in \mathbb{N}$ is the descriptor size in bytes, i.e., 16 bytes in our case):

$$d\_start = old\_d - \frac{((old\_d \times \mathcal{D}) \bmod \mathcal{C})}{\mathcal{D}} \qquad (6.4)$$

$$d\_end = \frac{cur\_d \times \mathcal{D} + \mathcal{C} - ((cur\_d \times \mathcal{D}) \bmod \mathcal{C})}{\mathcal{D}} \qquad (6.5)$$

For one sampling interval $\mathcal{A}_e^{TX_{reads}}$, $\mathcal{A}_e^{TX_{writes}}$, $\mathcal{A}_e^{RX_{reads}}$, and $\mathcal{A}_e^{RX_{writes}}$ are computed as follows:

$$\mathcal{A}_e^{TX_{reads}} = \sum_{n=1}^{cur\_d^{TX} - old\_d^{TX}} \left(1 + \frac{ba\_end_n^{TX} - ba\_start_n^{TX}}{\mathcal{C}} \right) \qquad (6.6)$$

It is necessary to add 1 memory read bus transaction for each transmit descriptor because of the corresponding descriptor fetch that is (according to our experiments) not optimized in terms of cache lines. This is handled differently when writing the descriptor done bit. In this case the ethernet controller tries to write as many descriptor done bits as possible. The maximum is four bits for one bus transaction.

$$\mathcal{A}_e^{TX_{writes}} = \frac{(d\_end^{TX} - d\_start^{TX}) \times \mathcal{D}}{\mathcal{C}} \qquad (6.7)$$

When receiving network packets, memory reads only occur due to receive descriptor fetching. We determined that the ethernet controller fetches four receive descriptors (equals to the cache line size) with one memory read bus transaction during our experiments. We use the indicator function with $N := \{n \in [old\_d^{RX}, cur\_d^{RX}] \mid (n \times \mathcal{D} \bmod \mathcal{C}) = 0\}$ in the following equation:

$$\mathcal{A}_e^{RX_{reads}} = \sum_{n=old\_d^{RX}}^{cur\_d^{RX}} \mathbb{1}_N(n) \qquad (6.8)$$

The number of expected bus transactions due to memory writes are as follows:

$$\mathcal{A}_e^{RX_{writes}} = \sum_{n=1}^{cur\_d^{RX} - old\_d^{RX}} \frac{ba\_end_n^{RX} - ba\_start_n^{RX}}{\mathcal{C}}$$
$$+ \frac{(d\_end^{RX} - d\_start^{RX}) \times \mathcal{D}}{\mathcal{C}} \qquad (6.9)$$

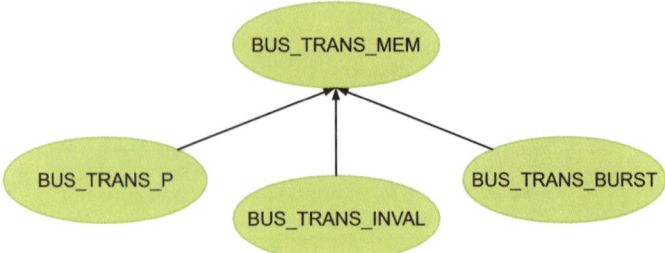

Fig. 6.4 BUS_TRANS event counter. The sum of BUS_TRANS_BURST, BUS_TRANS_P and BUS_TRANS_INVAL counts results in BUS_TRANS_MEM counts [71]

We expect that network packet data must be copied to the host memory and that corresponding descriptor done bits will be written to the descriptors in the host memory.

Exploiting Additional BUS_TRANS Events We verified Eq. 6.1 with further BUS_TRANS event counter that are basically subsets of the event BUS_TRANS_MEM, see Fig. 6.4. We determined that the event counter BUS_TRANS_P counts the memory reads of a peripheral and that the event counter BUS_TRANS_INVAL counts the memory writes of a peripheral. We used these counters in conjunction with THIS_AGENT and ALL_AGENTS name extensions as described in Sect. 5.2.1 to distinguish bus transactions caused by the host CPU and bus transactions caused by the peripheral. The event BUS_TRANS_BURST did not occur during our experiments. The number of bus transactions caused by the ethernet controller is computed according to Eq. 6.1 when the e1000e.ko driver function e1000_clean_tx_irq or e1000_clean_rx_irq is called. We enhanced BARM as introduced in Sect. 5.2.2 to consider $\mathcal{A}_e^{ETH}$ as described in this section.

6.2.2 Implementation Based on OpenSSL

OpenSSL is a popular software toolkit that implements cryptographic mechanisms such as the SSL/TLS protocol and the encoding/decoding of X.509 certificates. The toolkit provides the developer with shared libraries, i.e., libssl and libcrypto. The openssl command line tool also makes use of these libraries. Applications that require the cryptographic mechanisms provided by OpenSSL can use the libraries directly. Note, the implementation presented in this section is based on our [10] previous trusted channel implementation. Our modifications are based on TLS and TLS related *Request for Comments* (RFC) documents, i.e., RFC4366 and RFC4680. Hence, the modifications are compliant with the TLS specification.

The TLS handshake protocol used to negotiate a session key of a secure channel needs to be adapted to consider BARM's measurement results. Considering the measurement results during the handshake enables the peer to determine if the target platform is already attacked by DMA malware. This helps the peer to decide if the target platform is trustworthy. The peer can abort the handshake of the authentic reporting channel if the other endpoint is considered untrustworthy. Note, due to our trust model we consider the host CPU as a channel endpoint. Other computing environments including the network interface card do not belong to the endpoint. We use asymmetric cryptography mechanisms and certificates to authenticate endpoints. In the following paragraphs we describe the used key exchange and certificate. We also describe extensions for the TLS *Hello* messages. Extensions to the TLS protocol are considered by Dierks and Rescorla [38]. To transmit BARM measurement results (platform state data) additional handshake messages are required. We use *Supplemental Data* messages for this purpose.

Key Exchange Type Our implementation of the authentic report channel is based on an adapted version of the TLS Diffie-Hellman Ephemeral RSA (DHE-RSA) handshake.[1] That means, to authenticate endpoint data an RSA signing key pair is used. For the negotiation of the session key Diffie-Hellman values are used. The public Diffie-Hellman part that is transmitted to the peer is signed by the secret part of the RSA signing key pair.

Endpoint Certificate To authenticate the endpoints, certificates (see *cert* in Figs. 6.6 and 6.7) are exchanged during the TLS handshake. When using DHE-RSA, the certificates exchanged via *Certificate* messages contain the public part PK_{sign} of the signing key pair $K_{sign} := (SK_{sign}, PK_{sign})$. We have to ensure that the secret key SK_{sign} is only available to the endpoint. Our certificates include a BARM related identifier to *bind* the TLS-based authentic reporting channel to the endpoint. A certificate that includes a BARM identifier is issued by a trusted third party that vouches for a correct BARM installation on the target platform and that the secret key part SK_{sign} is only available on that endpoint. Hence, the certificate *cert* links the signing key K_{sign} to the endpoint that executes BARM. K_{sign} key pairs must be used to authenticate data sent by the client C and server S during the handshake. This eventually binds the transmitted platform state data to the authentic reporting channel. The trusted third party that vouches for the correct BARM installation and for the secret signing key part SK_{sign} could be the administrator who also runs the evaluation platform that receives platform state data (BARM measurements) from the target platform. The used certificate is actually a normal TLS certificate that includes the BARM related identifier. The certificate as well as the signing key pair K_{sign} are deployed together with BARM and are considered as long-lived.

---

[1] As described in prior work [10] other key exchange methods such as RSA and DH-RSA can also be used to implement a trusted channel based authentic reporting application.

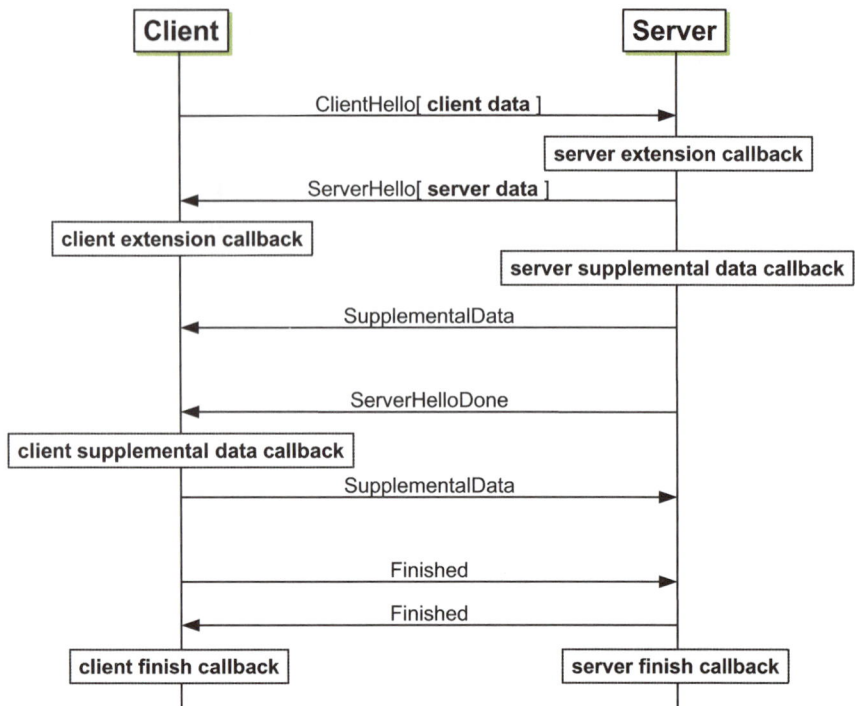

Fig. 6.5 TLS handshake considering hello extensions and supplemental data extensions. The *ClientHello* message contains `client data` and the *ServerHello* message contains `server data`. Additional *SupplementalData* messages contain `client supplemental data` and `server supplemental data`. Supplemental data is also considered as TLS extension. (based on [142])

Modifications to Hello Messages We use the *ClientHello* and *ServerHello* messages to negotiate the security parameters of the authentic reporting channel, see Fig. 6.6. The client platform *C* that runs BARM starts the adapted TLS client and sends the *ClientHello* message to the server platform *S*. The server replies with *ServerHello*. The *Hello* messages include the security parameters *sec_param* (see Sect. 6.1) of the corresponding peer, see Fig. 6.6. The security parameters determine which endpoint has to provide platform state data, i.e., BARM measurements. We use *Hello message extensions* [38] to exchange security parameters. Our OpenSSL-based implementation makes use of the *TLS Hello Extensions* as described in RFC4366 [14]. A patch for OpenSSL (0.9.8.x) implements the hello extensions, see Fig. 6.5.[2] The patch modifies code related to the library `libssl`. We use this patch for our authentic reporting channel application implementation.

---

[2] The TLS hello extensions and supplemental data patch can be found at http://openssl.6102. n7.nabble.com/PATCH-TLS-hello-extensions-and-supplemental-data-td38202.html [accessed 25 February 2014].

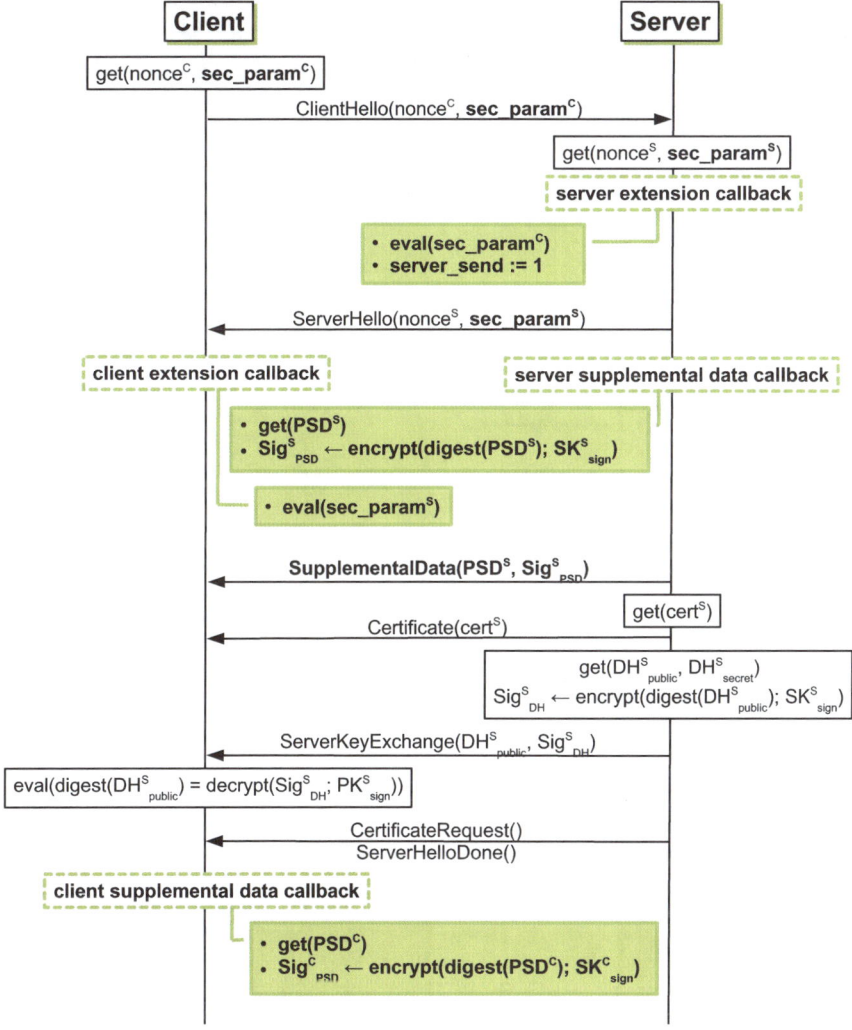

Fig. 6.6 Adapted TLS-DHE-RSA handshake for the authentic reporting channel (a). Modifications that were made to the TLS handshake are highlighted in bold text. The adapted handshake is continued in Fig. 6.7

The patch provides an interface that allows the developer to register new TLS extensions (see [142]). A TLS extension that is represented by the TLSEXT_GENE-RAL object transmits generic data. The application that uses TLS specifies the data format of the generic data. TLS extensions consist of a type, the data length, and

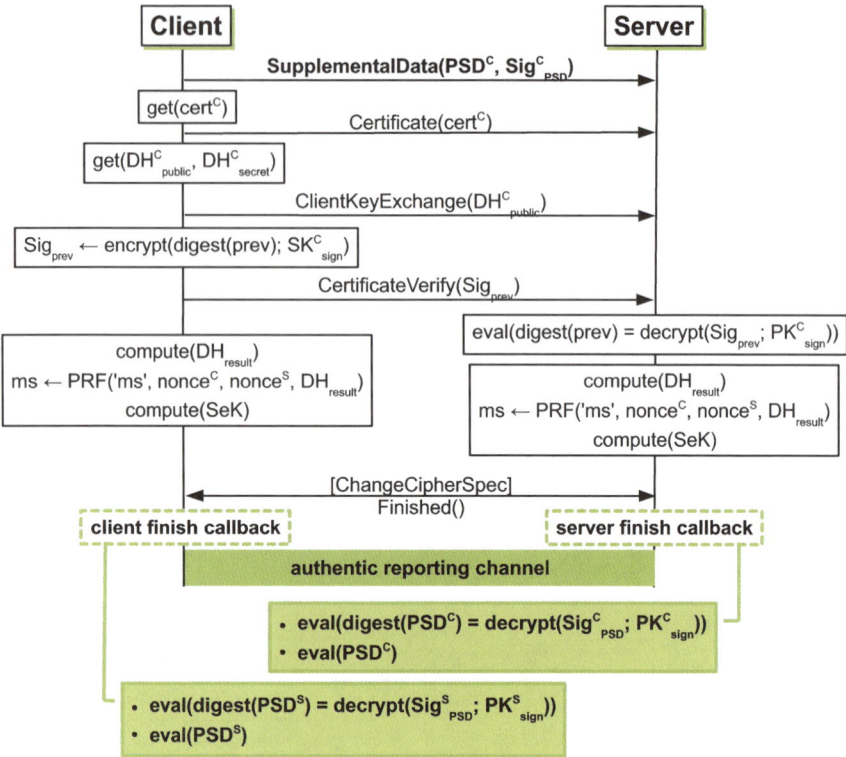

Fig. 6.7 Adapted TLS-DHE-RSA handshake for the authentic reporting channel (b). After the handshake has been finished the authentic reporting channel is used by BARM to transmit heartbeat messages in a regular interval to communicate platform state changes, i.e., to report a DMA malware based attack

the generic data (type-length-value format) as well as certain flags[3] and callback functions that implement the required extension logic. Callbacks (see Fig. 6.5) are only triggered on the peer that instantiated the corresponding TLSEXT_GENERAL object. The generic data that is transmitted via a *Hello* message is one generic datum. In our implementation the TLS extension that is exchanged via *Hello* messages (hello extension) is:

- ARCH_NEGOTIATION_EXT: This extension (EXT) for our authentic reporting channel (ARCH) is used to negotiate security parameters *sec_param*.

---

[3] The extension flags are client_required (the client will abort if the server ignores the extension where this flag has been set), server_send (the server will send the extension where this flag has been set), and received (internal use, e. g., to check duplicates).

Client as well as server register hello extensions (TLSEXT_GENERAL objects) if they want to handle them. If a peer receives a *Hello* message that contains the registered extension, the peer calls the corresponding extension callback, see Fig. 6.5.

Supplemental Data Messages for Platform State Data The client platform C as well as the server platform S can provide platform state data. We use so-called *SupplementalData* messages (see Fig. 6.5) as specified by the *Internet Engineering Task Force Networking Group* in RFC4680 [113] to transmit platform state data. The OpenSSL patch also implements *SupplementalData* messages for OpenSSL (0.9.8.x).[4] The details of the implementation of this patch are explained by Davide Vernizzi [142]. As described in RFC4680 supplemental data is also used to transmit generic data. The peer determines whether or not the generic data needs to be transmitted using hello extensions. The OpenSSL patch also enables us to define supplemental data extensions that we need for our authentic reporting channel. Supplemental data extensions also consist of a type, the generic data, the data length, and callback functions. Supplemental data transmitted using the *SupplementalData* message can be a stack of several generic data. In our implementation the extensions to exchange generic data via *SupplementalData* messages are:

- ARCH_SUPP_DATA_C_EXT: This extension is used to transmit the platform state data PSD^C (supplemental data) from the client C to the server S.
- ARCH_SUPP_DATA_S_EXT: This extension is used to transmit the platform state data PSD^S (supplemental data) from the server S to the client C.

The patched OpenSSL software handles generic data as presented in Fig. 6.5. Callback functions that also belong to the TLS extensions are called to process the generic data according to the required extension logic. Analogous to hello extensions, client and server have to register for supplemental data extensions that they want to handle via the corresponding supplemental data callbacks. Figures 6.6 and 6.7 depict how our generic data (hello extensions as well as supplemental data extensions) is handled using the callback functions during the adapted TLS handshake.

In our proof of concept implementation the generic data format used to exchange platform state data *PSD* via supplemental data is quite simple:

- barm_measurement: This data field contains the BARM measurement taken by the BARM Linux kernel module.
- Devices flag pair list: We use a devices flag pair list to communicate if a peripheral is attacking the target platform. The first flag represents if the corresponding device started to attack the host and, if so, the second flag states if the malicious device could be stopped. The devices flag pair list looks as follows:
  - (uhci_attack, uhci_disabled): This flag pair represents the UHCI controller.
  - (..._attack, ..._disabled): *[further devices]*

---

[4] See Footnote 2.

- (me_attack, me_disabled): This flag pair represents the manageability engine.

- *nonceSD*: *nonceSD* consists of the two elements:

  - *nonce^C* (client_random)
  - *nonce^S* (server_random)

The signature Sig_{PSD} on the platform state data *PSD* is also sent to the peer via the *SupplementalData* message, see Figs. 6.6 and 6.7. By doing so, the platform state data *PSD* is also bound to the corresponding secure channel. The *nonceSD* included in the supplemental data is compared with *nonce^C* and *nonce^S* (sent via the *Hello* messages) to guarantee freshness of the received platform state data *PSD*. To authenticate and to be able to check the integrity of platform state data *PSD*, we use the secret part SK_{sign} of the signing key pair to sign *PSD*. To be able to verify the signature each peer provides the certificate that contains the public key part PK_{sign} using the *Certificate* message directly after transmitting the *SupplementalData* message, see Figs. 6.6 and 6.7. The BARM measurement results that are also part of the supplemental data are evaluated to derive the trustworthiness of the peer. Depending on the derived trustworthiness the local platform takes measures according to the local security policy.

Session Key Computation The session key *SeK* is computed on both peers as usual. Since we use DHE-RSA, the secure channel that uses *SeK* is eventually linked to the endpoints (host CPUs). The exchanged DH parts are signed using the secret part of K_{sign} (SK_{sign}) that links the DH values to K_{sign}. The signing key pair K_{sign} is bound to exactly one endpoint due to the certificate issued by the trusted third party that vouches for the fact that SK_{sign} is only available on the endpoint. Hence, the session key is also bound to the endpoint.

Heartbeat Messages After the handshake has been completed, BARM uses the negotiated channel to send heartbeat messages in a regular interval to the external administrator platform. These messages contain the current BARM measurement and the devices flag pair list in a similar *PSD* format that has been used during the handshake. Only *nonceSD* is missing. The regular heartbeat messages are used by BARM to report platform state changes, i.e., a DMA malware based attack. If the external platform does not receive heartbeat messages anymore we assume that the NIC tried to attack the host platform and BARM was able to successfully stop the attack. It is also possible that malware that is executed on the NIC blocks the heartbeat messages. If so, the attack is also revealed.

6.3 Evaluation

We use the same platform and basic evaluation configuration as described in Sect. 5.3 to evaluate the enhanced BARM. Please note, only the client platform must transmit platform state data.

6.3.1 Expected Bus Activity Validation

To validate Eq. 6.1 we conducted different tests. The evaluation results are depicted in Fig. 6.8. The results reveal larger fluctuations in BARM measurement results when the `ping`, `scp` and `wget` command cause network traffic. Table 6.1 provides information on the cause of the larger fluctuations. The table presents BARM measurements that were taken during the download of a 1 GB file using the `wget` command. The applied sampling interval was 32 ms. The table depicts that a larger positive discrepancy (see BARM sample 125,924: 13 bus transactions) is followed by a larger negative discrepancy (see BARM sample 125,925: −12 bus transactions). We assume that a positive discrepancy occurs when network packets were already copied to the host memory by the ethernet controller, but BARM was unable to evaluate the corresponding receive descriptors in the current sampling interval. These descriptors are available in the next sampling interval. Hence, BARM evaluates the descriptors in the next interval, which in turn results in a negative discrepancy. BARM subtracts expected bus transactions from the measured transactions that were actually measured in the last sampling interval.

As depicted in Table 6.1, the positive and negative values compensate one another. Thus, the fluctuation can be minimized by simply adding positive and negative

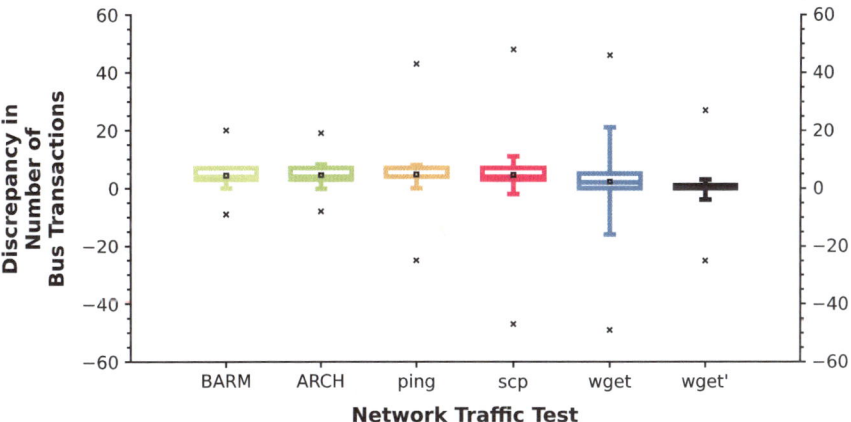

Fig. 6.8 Expected bus activity evaluation with network traffic. We evaluated the expected bus activity for six different test cases. The discrepancy is visualized in the form of boxplots as known from Fig. 5.6. In the first case (BARM) we only run the enhanced BARM and in the second case we run the enhanced BARM together with the OpenSSL-based authentic reporting channel. We took 100 BARM measurements in both cases. BARM and the authentic reporting channel are also active in the remaining test cases (ping, scp, wget, wget'). We executed the `ping` command with a 1,000 bytes payload 100 times (ping). In the case of scp we copied a 100 MB file from an external platform to our target platform 100 times. In the wget case we downloaded a 1 GB file from http://download.thinkbroadband.com/1GB.zip [accessed 25 February 2014] using the `wget` command. We applied a BARM sampling interval of 32 ms for all test cases except the last one (wget'). The boxplot for the wget' case represents the result when using a sampling interval of 1,024 ms during a `wget` download of a 1 GB file

Table 6.1 BARM measurement values revealing fluctuations

BARM sampling number	BARM measurement value	BARM sampling number	BARM measurement value
125,912	5	125,944	2
125,913	2	125,945	25
125,914	3	125,946	−48
125,915	2	125,947	28
125,916	3	125,948	0
125,917	0	125,949	3
125,918	0	125,950	5
125,919	1	125,951	1
125,920	2	125,952	13
125,921	−17	125,953	−21
125,922	22	125,954	5
125,923	1	125,955	2
125,924	13	125,956	2
125,925	−12	125,957	−1
125,926	−15	125,958	4
125,927	22	125,959	3
125,928	5	125,960	−21
125,929	0	125,961	25
125,930	−2	125,962	2
125,931	5	125,963	−2
125,932	2	125,964	3
125,933	5	125,965	2
125,934	0	125,966	5
125,935	−2	125,967	2
125,936	9	125,968	−1
125,937	2	125,969	2
125,938	3	125,970	2
125,939	−2	125,971	6
125,940	8	125,972	0
125,941	−3	125,973	1
125,942	0	125,974	2
125,943	5	125,975	3

The sampling numbers and the corresponding measurement values taken are from the measurement log that was taken when downloading a 1 GB file from http://download.thinkbroadband.com/1GB. zip [accessed 25 February 2014]. The BARM sampling interval was 32 ms

BARM measurement values. As presented in Table 6.1, a pair of positive and negative measurement values can also occur the other way around (see BARM samples 125,926 and 125,927, for example). This means that the negative value is determined before the positive value. We assume that this occurs when BARM has already analyzed transmit descriptors when the corresponding packets were not copied by the ethernet controller yet. Hence, BARM already subtracts the expected bus transactions from the measured ones before they are actually measured. The transactions are measured in the next sampling interval that results in a larger positive discrepancy.

We examined the described behavior with two sampling intervals when using the `wget` command to download a 1 GB file. As depicted in Fig. 6.8, the fluctuations are larger when using `wget` with a sampling interval of 32 ms (see **wget**) compared to a sampling interval of 1,024 ms (see **wget'**).

6.3.2 Network Performance Overhead Evaluation

We conducted a network benchmark to reveal the network performance overhead that is caused by the enhanced BARM version. The enhanced BARM version permanently sends heartbeat messages. The results are presented in Fig. 6.9. The results in Fig. 6.9 reveal a relative performance overhead of approximately 4.5 % when sending the heartbeat message every 32 ms. This interval length corresponds the BARM sampling interval. It is not necessary to use the same interval for reporting as for BARM measurement sampling due to the heartbeat message format that we use to transmit platform state data. The devices flag pair list represents a history of malicious peripherals. Hence, the network performance overhead for 32 ms sampling and

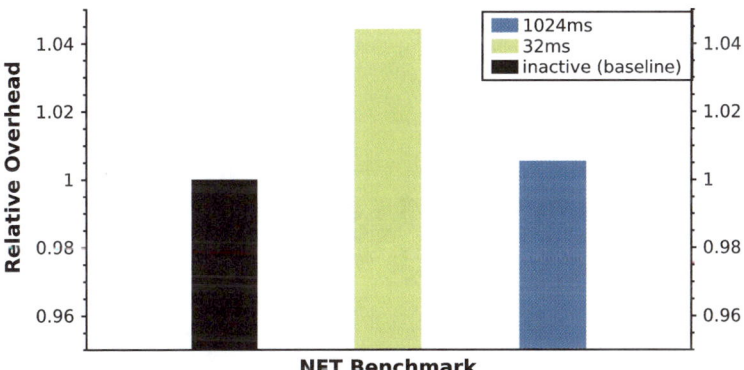

Fig. 6.9 Relative performance overhead for different reporting intervals and constant sampling interval. The figure compares the results of three measurement series. The first measurement series (inactive) represents the baseline. Inactive means that BARM was not running and no heartbeat messages were sent. A bar in the figure represents the mean of 100 measurements, see also Sect. 5.3.2. We measured the clock cycles (with time stamp counters) that are needed to copy a 100 MB file from an external platform with the `scp` command. Measurements were taken for a reporting interval of 32 ms and for a reporting interval of 1,024 ms. In both cases we used the same BARM sampling interval of 32 ms. The relative performance overhead when sending a heartbeat message every 32 ms is approximately 4.5 %. The overhead is only approximately 0.5 % when sending the message every 1,024 ms

reporting interval can be avoided. The only requirement is that the sampling interval
is less or equal than the reporting interval.

6.3.3 Test with DAGGER

We repeated the DMA malware DAGGER test (see Sect. 5.3.3) with our enhanced
bus agent runtime monitor BARM. The results are summarized in Figs. 6.10, 6.11
and 6.12. We attacked the target platform at an arbitrary point in time during runtime.
Figure 6.10 confirms that the enhanced BARM could reveal the DMA attack as well
as stop the malicious peripheral. The excerpt from the log in Figs. 6.11 and 6.12
belong to the same experiment that was the basis for Fig. 6.10.

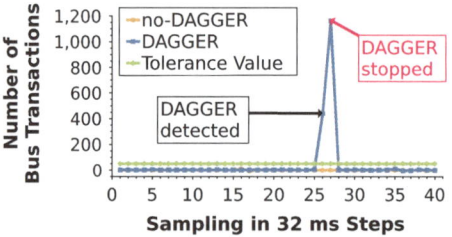

Fig. 6.10 Evaluating enhanced BARM at an arbitrary point during runtime with the authentic
reporting channel. The conducted experiment is similar to the experiment presented in Sect. 5.3.3.
BARM's sampling interval was 32 ms and the tolerance value was 50 bus transactions. This time
BARM considers the ethernet controller as an additional bus master that allowed us to start our
authentic reporting channel. Heartbeat messages were sent every 32 ms. The figure compares three
curves, i.e., the tolerance value $\mathcal{T}$, BARM's measurement results without any attack, and BARM's
measurement results with a DAGGER attack

```
[ 5698.721004] BARM: 4
[ 5698.753004] BARM: 2
[ 5698.785014] BARM: 1
[ 5698.817005] BARM: 1
[ 5698.849005] BARM: 4
[ 5698.885892] BARM: 441
[ 5698.885894] BARM: DMA ATTACK DETECTED! Checking UHCI controller ...
[ 5698.913042] BARM: 1163
[ 5698.913043] BARM: UHCI controller is not attacking. Checking Intel's Manageability Engine ...
[ 5698.945006] BARM: 1
[ 5698.945008] BARM: The ME controller has attacked the platform. ME controller stopped!
[ 5698.977007] BARM: 2
[ 5699.009071] BARM: 4
[ 5699.041005] BARM: 2
[ 5699.073031] BARM: 0
[ 5699.105005] BARM: 2
[ 5699.137049] BARM: 2
```

Fig. 6.11 BARM authentic reporting channel—client side. The figure presents a part of BARM's
log output. BARM is deployed on the target platform, i.e., the client. The log output demonstrates
that BARM revealed a DMA attack and that BARM was able to stop the malicious peripheral

```
SSL_accept:before/accept initialization
<<< TLS 1.0 Handshake [length 005e], ClientHello
BARM Authentic Channel: ssl_tlsext_test_server_ext_cb called.
SSL_accept:SSLv3 read client hello A
>>> TLS 1.0 Handshake [length 0035], ServerHello
SSL_accept:SSLv3 write server hello A
BARM Authentic Channel: ssl_tlsext_test_server_supp_data_cb called.
        BARM Trusted Channel: No server measurement result required.
>>> TLS 1.0 Handshake [length 000e]???
SSL_accept:unknown state
>>> TLS 1.0 Handshake [length 0c5f], Certificate
SSL_accept:SSLv3 write certificate A
>>> TLS 1.0 Handshake [length 028d], ServerKeyExchange
SSL_accept:SSLv3 write key exchange A
>>> TLS 1.0 Handshake [length 00a8], CertificateRequest
SSL_accept:SSLv3 write certificate request A
SSL_accept:SSLv3 flush data
<<< TLS 1.0 Handshake [length 0270]???
SSL_accept:unknown state
<<< TLS 1.0 Handshake [length 0c5f], Certificate
SSL_accept:SSLv3 read client certificate A
<<< TLS 1.0 Handshake [length 0046], ClientKeyExchange
SSL_accept:SSLv3 read client key exchange A
<<< TLS 1.0 Handshake [length 0206], CertificateVerify
SSL_accept:SSLv3 read certificate verify A
<<< TLS 1.0 ChangeCipherSpec [length 0001]
<<< TLS 1.0 Handshake [length 0010], Finished
SSL_accept:SSLv3 read finished A
>>> TLS 1.0 Handshake [length 065a]???
SSL_accept:SSLv3 write session ticket A
>>> TLS 1.0 ChangeCipherSpec [length 0001]
SSL_accept:SSLv3 write change cipher spec A
>>> TLS 1.0 Handshake [length 0010], Finished
SSL_accept:SSLv3 write finished A
SSL_accept:SSLv3 flush data
BARM Authentic Channel: ssl_tlsext_test_server_finish_cb called.
[...]
00000004
00000002
00000001
00000001
00000004
00000441
00001163
00000001
00000002
00000004
00000002
00000000
[...]
```

Fig. 6.12 BARM authentic reporting channel—server side. The figure depicts the log output of the adapted OpenSSL server. The log consists of the TLS handshake messages, callback call messages, and received BARM measurements. The measurement values are the same as presented in Fig. 6.11. The BARM instance that is deployed on the client side was able to stop the attack. In this example, the local security policy tolerates the stopped attack. Alternatively, the server could have torn down the channel when the server received the BARM measurement of 441 bus transactions. The server was also configured with $\mathcal{T} = 50$ bus transactions

6.4 Security Considerations

In this section we informally evaluate the security requirements that we introduced in the beginning of this chapter. A formal proof is outside the scope of this work. Many research related to security proofs of the TLS protocols have been published in the past. An overview is presented by Kohlweiss et al. [78]. The research also considers multiple TLS variants. We assume that our TLS-based channel can also be formally proven. However, the focus of this chapter is the enhanced BARM that considers the network interface card. Hence, we review the extent to which our enhanced BARM fulfills the requirements for a secure channel (R1), binding of BARM measurements to the secure channel (R2), and privacy (R3).

- **R1—Secure channel properties**: Due to the applied TLS protocol the secure channel properties confidentiality, integrity, authenticity as well as freshness are ensured for the communication channel. Due to the enhanced BARM these properties are also ensured on the endpoint, i.e., the host CPU. Given that the attacker has to search for valuable data, BARM ensures the integrity and the confidentiality of data that is present in the main memory. The attacker could merely randomly write to or read from the main memory without searching for valuable data. The attacker also needs to search for nonces, key material or the session key SeK as well as the private part of the signing key pair SK_{sign} to attack the communication session. Hence, the enhanced BARM also takes care of the properties authenticity and freshness on the endpoint due to the detection of additional bus traffic when the attacker searches in the main memory.

  The attacker can only conduct a MitM attack if the attacker is able to steal private key material or the session key via DMA. Scanning the memory for this data will be detected by BARM. BARM can also identify the malicious device. Hence, the access to the main memory can be prevented. Note, the host CPU could enforce the attacker to cause more bus transactions by storing parts of the sensitive data in processor registers. This technique was proposed in related work, see Sect. 3.2.6. This will not protect the sensitive data, since DMA attacks can be used to dump the content of processor registers into the main memory. However, such an attack will cause more bus activity, which will also be detected by BARM.

  The attacker could attempt to modify BARM measurements. To do so, the attacker could try to find the variables in the main memory where BARM stores the values of the performance monitoring units that we exploit to reveal DMA attack. However, the DMA-based search would be revealed by BARM. Alternatively, the attacker could try to modify the host CPU registers that correspond to the performance monitoring units used by BARM. The attacker has no direct access to host CPU registers. However, the attacker needs to find a memory area to store host CPU instructions that modify the performance monitoring processor registers. It is required that the host CPU will sooner or later consider the memory area, which contains the malicious instructions. Again, the attacker has to search for such an area via DMA and this DMA-based search will be revealed by BARM.

- **R2—Binding of BARM measurements to the secure channel**: Authenticity of an endpoint is ensured by providing the certificate *cert* that includes the BARM identifier and the public key part of the signing key pair PK_{sign}. The certificate is signed by a trusted party. Two factors ensure that the BARM measurements are bound to the channel. First, the BARM measurement that is transmitted during the handshake is signed using the endpoint's secret part of the signing key SK_{sign}. Second, the exchanged DH values that are used for the session key computation are also signed with SK_{sign}. Hence, not only the first transmitted BARM measurement as well as the DH values are bound to the channel endpoint, but also the session key *SeK* that eventually establishes the secure communication channel for authentic state reporting. This means that every heartbeat message is also bound to the channel endpoint. These messages are only transmitted in encrypted form via the channel that is protected by *SeK*.

  The endpoint's authenticity also prevents a relay attack where the attacker could send a request to a third platform to sign platform state data *PSD* that includes a BARM measurement value that is less than 50 bus transactions. The third platform has no access to SK_{sign} of the target platform. That means, we can exclude that the attacker is able to conduct a relay attack. Alternatively, the attacker could try to forge a *PSD* signature. To do so, the attacker requires SK_{sign} that is present in the main memory. Again, when the attacker searches for SK_{sign} via DMA, BARM will reveal this attack and the memory access will be prevented. Hence, we can conclude that the attacker is unable to forge digital signatures.

- **R3—Privacy**: The only sensitive data that is transmitted unencrypted is the first BARM measurement value that is sent to the peer during the handshake. While a compromised network interface card could be used to intercept this value, it is unlikely that this first measurement value is of use for an attacker. It is independent of further measurement values, which are required to identify when BARM determines $-T$ bus transactions. Hence, we can conclude that our authentic reporting channel adheres to the least information paradigm.

6.5 Chapter Summary

In this chapter we developed, implemented, and evaluated an authentic reporting channel application for BARM. This channel is based on the secure channel protocol TLS. We modified the TLS protocol to consider BARM measurements during the handshake as well as during the rest of the communication session. Our modifications are based on TLS extensions. This means that our channel is compliant with the TLS specification. Furthermore, the implementation of our reporting channel fulfills the security requirements (host CPU endpoint authenticity and channel binding) that we defined for the DMA malware scenario. Without the fulfillment of these requirements malware executed on the network interface card is a threat for an authentic communication with an external platform.

Our channel is an application for our bus agent runtime monitor if platform state change reporting is required by a communication partner. The authentic reporting channel transmits the state changes to the peer. We confirmed BARM's effectiveness and efficiency with our DMA malware DAGGER in conjunction with the implemented reporting channel. Previous work that is related to authentic platform state reporting assumed the presence of an efficient runtime monitor. However, the corresponding proof of concept implementations presented in previous work did not include such a monitor. Furthermore, previous work did also not consider the DMA malware scenario.

We can also conclude that BARM can handle more complex bus masters. We demonstrated that BARM can not only handle the host CPU, the UHCI controller, etc., but also the ethernet controller. To integrate the ethernet controller into BARM's detection model we had to analyze the controller with regards to memory read and write accesses. We were able to distinguish read and write accesses by exploiting additional performance monitoring unit configurations. However, to eventually determine the number of bus transactions that are caused by the ethernet controller we had to introduce a new parameter. This new parameter is the cache line size. According to our evaluation, BARM measurement fluctuations are minimally higher as compared to the BARM version that does not consider the ethernet controller. Nonetheless, the fluctuations are still in the range of $\mathcal{T} = +/-50$ bus transactions. Our empirical measurements revealed that the performance overhead of the authentic reporting application is negligible if heartbeat messages are sent approximately every second. The reporting interval can be greater than BARM's sampling interval. The loss of DMA malware attack information is prevented by including an attack history in the heartbeat messages.

Chapter 7
Conclusions and Future Work

Logic is a systematic method of coming to the wrong conclusion with confidence.

Manly's Maxim,
Murphy's Law Collection

The compromise of computer platform peripherals to attack the host platform memory currently represents the peak of the rootkit evolution. This thesis presents a study on computer platform attacks that exploit such rootkit techniques. Platform peripherals are well-suited for hiding malicious code to attack the host platform. The peripherals consist of an isolated execution environment with a dedicated processor, dedicated memory and direct access to the host memory. Prior to this work, attacks originating from malware that exploits direct memory access (DMA) were considered to be *invisible* to the host CPU. Security software such as state-of-the-art anti-virus software does not consider the isolated execution environments. However, this thesis demonstrates that *the host CPU is able to detect attacks that exploit DMA. This enables the host CPU to mitigate such attacks.*

Nowadays, peripherals such as management controllers and network interface cards (NIC) are present in almost every computing device. Server systems, desktop systems, laptops, tablets, and even mobile phones use dedicated controllers to offload work from the host CPU. Although it is a resource intensive task to infiltrate such a peripheral, these environments remain attractive in terms of stealthiness. The DMA mechanism is the basis for attacking the host memory. Hence, we call peripheral-based attack code, that exploits direct memory access, DMA malware. With DMA malware an attacker can read from and write to the host main memory in a stealthy manner. The adversary can have access to all data present in the main memory. Therefore, the attacker can steal sensitive data such as cryptographic keys, passwords, internet banking credentials, open documents, as well as all user input. The adversary can also insert data into the main memory to implement a kernel backdoor. However, this lowers the probability of a successful stealthy attack, since host software could theoretically detect the malicious modification to the host memory. Conversely, the attacker can also attack the host detection software via DMA to prevent the detection.

© Springer International Publishing Switzerland 2015
P. Stewin, *Detecting Peripheral-based Attacks on the Host Memory*,
T-Labs Series in Telecommunication Services, DOI 10.1007/978-3-319-13515-1_7

In this work we developed and analyzed a DMA malware proof of concept. The malware is executed on an isolated execution environment whose inner workings is inaccessible by the host. The goal of this thesis was to demonstrate that the host CPU can defend itself against DMA malware even if the host CPU is unable to access the inner workings of the suspected peripheral. The peripheral that we used for our malware proof of concept is *Intel's Manageability Engine* (Intel ME). Amongst other things, Intel utilizes the ME environment to implement a web server, which provides system administrators with remote device management capabilities. Administrators can recover the host OS even when the platform does not boot up anymore, e.g., due to OS kernel integrity corruptions. Intel applied protection mechanisms to ensure that ME features cannot be exploited to attack the host. However, this protection also ensures that, e.g., anti-virus software is incapable of evaluating the ME environment. Conversely, an attacker capable of infiltrating the ME environment, e.g., with a zero-day exploit, also benefits from this protection.

Our malware proof of concept is a *Direct memory Access based keystroke code loGGER*. DAGGER demonstrates that it is possible to implement stealthy malware in terms of detection capabilities of the host CPU. The attack code is executed on the dedicated ME processor. Thus, the keystroke code logger does not result in a measurable performance overhead for the host. Our malware is also capable of capturing short living data such as keystroke codes. We exploit the isolated out-of-band network feature of the ME environment to exfiltrate private data such as the captured keystrokes to an external platform. This network feature is also invisible to the host. Our analysis of DAGGER revealed that DMA malware must search for valuable data in the host memory. The process of searching the host memory results in additional bus activity, which increases if memory address randomization mechanisms are used or if the secret data remains in the CPU cache or CPU registers. We also determined that parallel memory access requests of different devices are arbitrated by the memory controller hub. This led to the assumption that the arbiter could cause DMA side effects that can be utilized to detect DMA malware. We confirmed this assumption by conducting a memory stress test. With this experiment we demonstrated a reliable measurable DMA side effect. Our measurements consider precise timing based on host CPU clock cycles in conjunction with performance counters.

We continued this research with the goal of developing a DMA malware detector. We analyzed the host CPU's performance counters. Finally, our investigation resulted in a performance counter configuration that is capable of distinguishing legitimate and illegitimate memory bus transactions. We modeled the expected bus activity of the host operating system and compared it to the measured bus activity. To model the expected bus activity we use information that is available on the host CPU in the operating system kernel.

We implemented our model and measurement mechanism in the form of an operating system kernel module that we call *Bus Agent Runtime Monitor*, in short BARM. BARM is a runtime monitor that also considers transient attacks. Our monitor causes only negligible performance overhead. BARM does not require any firmware and hardware modifications. Our runtime monitor also does not require any access to the inner workings of potentially compromised peripherals. After evaluating our proof of concept implementation of BARM, we can conclude that the host CPU is able to detect and halt DMA malware. Our evaluation also revealed minimal BARM measurement fluctuations. Such fluctuations can occur when working with performance counters, which we used for our proof of concept implementation. We overcame this issue by introducing a tolerance value. The tolerance value is an empirical value that represents tolerable bus transactions, i.e., in our case the tolerance value is 50 bus transactions. We demonstrated that DMA malware causes significantly higher bus activity when searching for valuable data in the host runtime memory.

However, the tolerance value also demonstrates a limitation of the current BARM implementation. Theoretically, an adversary could hide up to $2T$ bus transactions per sampling interval. This can only work if the adversary is able to predict the exact points in time when BARM determines $-T$ bus transactions. Conversely, this would also result in a slower search phase that could be exploited by the host CPU to protect the target data in the host memory. Another limitation was a possible MitM attack conducted by the ethernet controller. We mitigated such MitM attacks by implementing an authentic reporting channel. The authentic channel for reporting the platform state was another goal of this work. The platform state contains DMA malware in our scenario. BARM delivers authentic measurement results to an external platform.

A secure channel protocol such as TLS is insufficient in our scenario. We adjusted the TLS protocol to meet the requirements of authentic platform state reporting. It is important to consider the NIC as well, since it could potentially modify or block BARM packets. To avoid detection the NIC could also implement a *Man-in-the-Middle* (MitM) attack by relaying benign BARM measurements of another platform to the platform that wants to evaluate the state of the target platform. Another example is to steal secret key material that is present in the host runtime memory via DMA. To eliminate these issues, the communication channel is bound to the actual endpoint, i.e., the host CPU. BARM digitally signs the bus activity measurements and ensures that the private key as well as the session key of the communication channel are protected from DMA malware. To implement a proof of concept authentic platform state reporting application we had to enhance BARM to consider legitimate memory bus activity of the NIC. The evaluation of our channel application confirms that the NIC can be reliably considered by BARM.

Future Work Although we can conclude that the host CPU is able to defend itself against DMA malware using BARM there are still some tasks left for future work. First of all, it would be interesting to evaluate the idea behind BARM on non-Intel hardware. Other architectures such as ARM also provide hardware performance counters. Platforms that are based on ARM also work with peripherals that are potential hosts for DMA malware. This is particularly interesting when considering that such platforms make extensive use of SoCs (*System on a Chip*) in their designs. Hence, peripherals within the same device package or die can be used to implement system backdoors.

It is also quite interesting to investigate if the timing-based DMA side effect can also be exploited to implement a reliable detection tool. This can be useful for architectures that do not support performance counters. The BARM implementation should also consider other peripherals as well. From our point of view it is more important to integrate additional peripherals in BARM's detection model. It is also possible to eliminate fluctuations in BARM's measurements. It is important to note that the integration of DMA-based devices is a resource intensive task. Therefore, a follow-up research project should examine to which extent this process can be automated.

References

1. Doug Abbott. *PCI Bus Demystified*. Demystifying Technology Series. Elsevier Science, 2004.
2. Darren Abramson, Jeff Jackson, Sridhar Muthrasanallur, Gil Neiger, Greg Regnier, Rajesh Sankaran, Ioannis Schoinas, Rich Uhlig, Balaji Vembu, and John Wiegert. Intel Virtualization Technology for Directed I/O. *Intel Technology Journal*, 10(3):179–192, August 2006.
3. Grace Agnew. *Digital Rights Management: A Librarian's Guide to Technology and Practise*. Chandos Information Professional Series. Chandos Pub., 2008.
4. Raja Naeem Akram, Konstantinos Markantonakis, and Keith Mayes. Application-binding Protocol in the User Centric Smart Card Ownership Model. In *Proceedings of the 16th Australasian Conference on Information Security and Privacy*, ACISP'11, pages 208–225, Berlin, Heidelberg, 2011. Springer-Verlag.
5. Raja Naeem Akram, Konstantinos Markantonakis, and Keith Mayes. A Privacy Preserving Application Acquisition Protocol. In Geyong Min, Yulei Wu, Lei (Chris) Liu, Xiaolong Jin, Stephen A. Jarvis, and Ahmed Yassin Al-Dubai, editors, *TrustCom*, pages 383–392. IEEE Computer Society, 2012.
6. Don Anderson. *FireWire System Architecture: IEEE 1394a*. PC System Architecture Series. Addison Wesley, 1999.
7. Don Anderson. *SATA Storage Technology*. MindShare Technology Series. MindShare Press, 2007.
8. Don Anderson and Dave Dzatko. *Universal Serial Bus System Architecture*. PC System Architecture Series. Addison Wesley, 2001.
9. Don Anderson and Tom Shanley. *PCI System Architecture*. PC System Architecture Series. Addison Wesley, 1999.
10. Frederik Armknecht, Yacine Gasmi, Ahmad-Reza Sadeghi, Patrick Stewin, Martin Unger, Gianluca Ramunno, and Davide Vernizzi. An Efficient Implementation of Trusted Channels based on OpenSSL. In *Proceedings of the 3rd ACM Workshop on Scalable Trusted Computing*, STC'08, pages 41–50, New York, NY, USA, 2008. ACM.
11. Damien Aumaitre and Christophe Devine. Subverting Windows 7 x64 Kernel with DMA Attacks. Sogeti ESEC Lab: http://esec-lab.sogeti.com/dotclear/public/publications/10-hitbamsterdam-dmaattacks.pdf [accessed 25 February 2014], July 2010.
12. M. Baugher, D. McGrew, M. Naslund, E. Carrara, and K. Norrman. The Secure Real-time Transport Protocol (SRTP). The Internet Engineering Task Force: http://tools.ietf.org/html/rfc3711 [accessed 25 February 2014], March 2004. RFC3711.
13. Muli Ben-Yehuda, Jimi Xenidis, Michal Ostrowski, Karl Rister, Alexis Bruemmer, and Leendert van Doorn. The Price of Safety: Evaluating IOMMU Performance. In *OLS'07: The 2007 Ottawa Linux Symposium*, pages 9–20, July 2007.

© Springer International Publishing Switzerland 2015   99
P. Stewin, *Detecting Peripheral-based Attacks on the Host Memory*,
T-Labs Series in Telecommunication Services, DOI 10.1007/978-3-319-13515-1

14. S. Blake-Wilson, M. Nystrom, D. Hopwood, J. Mikkelsen, and T. Wright. Transport Layer Security (TLS) Extensions. The Internet Engineering Task Force: http://www.ietf.org/rfc/rfc4366.txt [accessed 25 February 2014], April 2006. RFC4366.
15. Erik-Oliver Blass and William Robertson. TRESOR-HUNT: Attacking CPU-bound Encryption. In *Proceedings of the 28th Annual Computer Security Applications Conference*, ACSAC'12, pages 71–78, New York, NY, USA, 2012. ACM.
16. Bill Blunden. *The Rootkit Arsenal: Escape And Evasion In The Dark Corners Of The System.* Jones & Bartlett Learning, 2012.
17. Adam Boileau. Hit by a Bus: Physical Access Attacks with Firewire. Security-Assessment. com: http://www.security-assessment.com/files/presentations/ab_firewire_rux2k6-final.pdf [accessed 25 February 2014], October 2006. Ruxcon 2006.
18. Rory Breuk and Albert Spruyt. Integrating DMA Attacks in Exploitation Frameworks. Homepage of Cees de Laat: http://www.delaat.net/rp/2011-2012/p14/report.pdf [accessed 25 February 2024], February 2012.
19. Rory Breuk and Albert Spruyt. Integrating DMA Attacks in Metasploit. Sebug: http://sebug. net/paper/Meeting-Documents/hitbsecconf2012ams/D2%20SIGINT%20-%20Rory%20Breuk%20and%20Albert%20Spruyt%20-%20Integrating%20DMA%20Attacks%20in%20Metasploit.pdf [accessed 25 February 2014], May 2012.
20. Ernie Brickell, Jan Camenisch, and Liqun Chen. Direct Anonymous Attestation. In *Proceedings of the 11th ACM Conference on Computer and Communications Security*, CCS'04, pages 132–145, New York, NY, USA, 2004. ACM.
21. Jonathan Brossard. Hardware Backdooring is Pratical. Toucan System: http://www.toucansystem.com/research/blackhat2012_brossard_hardware_backdooring.pdf [accessed 25 February 2014], 2012.
22. Jonathan Brossard. Hardware Backdooring is Pratical. Black Hat USA 2012: https://media.blackhat.com/bh-us-12/Briefings/Brossard/BH_US_12_Brossard_Backdoor_Hacking_Slides.pdf [accessed 25 February 2014], 2012.
23. William Buchanan. *Computer Busses.* Electronics & Electrical. Taylor & Francis, 2010.
24. Ravi Budruk, Tom Shanley, and Don Anderson. *PCI Express System Architecture.* The PC System Architecture Series. Addison Wesley, Pearson Education, July 2010. MindShare Inc.
25. Yuriy Bulygin. Chipset based Approach to Detect Virtualization Malware. hakim.ws: http://www.hakim.ws/BHUSA08/speakers/Bulygin_Detection_of_Rootkits/bh-us-08-bulygin_Chip_Based_Approach_to_Detect_Rootkits.pdf [accessed 25 February 2014], 2008.
26. John Butterworth, Corey Kallenberg, Xeno Kovah, and Amy Herzog. Problems with the Static Root of Trust for Measurement. Black Hat: https://media.blackhat.com/us-13/US-13-Butterworth-BIOS-Security-WP.pdf [accessed 25 February 2014], 2013. Presented at Black Hat, Slides: https://media.blackhat.com/us-13/US-13-Butterworth-BIOS-Security-Slides.pdf [accessed 25 February 2014].
27. Emanuele Cesena, Hans Löhr, Gianluca Ramunno, Ahmad-Reza Sadeghi, and Davide Vernizzi. Anonymous Authentication with TLS and DAA. In Alessandro Acquisti, Sean W. Smith, and Ahmad-Reza Sadeghi, editors, *Trust and Trustworthy Computing*, volume 6101 of *Lecture Notes in Computer Science*, pages 47–62. Springer, Berlin Heidelberg, 2010.
28. Xiaolin Chang, Ying Qin, Zhi Chen, and Bin Xing. ZRTP-based Trusted Transmission of VoIP Traffic and Formal Verification. In *Proceedings of the 2012 Fourth International Conference on Multimedia Information Networking and Security*, MINES'12, pages 560–563, Washington, DC, USA, 2012. IEEE Computer Society.
29. Song Cheng, Liu Bing, Xin Yang, Yang Yixian, Li Zhongxian, and Yin Han. A Security-enhanced Remote Platform Integrity Attestation Scheme. In *Proceedings of the 5th International Conference on Wireless Communications, Networking and Mobile Computing*, WiCOM'09, pages 4420–4423, Piscataway, NJ, USA, 2009. IEEE Press.
30. David Chess, Joan Dyer, Noamaru Itoi, Jeff Kravitz, Elaine Palmer, Ronald Perez, and Sean Smith. Using Trusted Co-servers to Enhance Security of Web Interaction. United States Patent 7,194,759: http://www.freepatentsonline.com/7194759.html [accessed 25 February 2014], March 2007.

31. Jonathan Corbet, Alessandro Rubini, and Greg Kroah-Hartman. *Linux Device Drivers, 3rd Edition*. O'Reilly Media Inc, 2005.
32. Rob Crooke. Accelerating Innovation in the Desktop. Intel Corporation: http://download.intel.com/pressroom/kits/events/computex2009/Crooke_Computex_presentation.pdf [accessed 25 February 2014], April 2009.
33. Francis M. David, Ellick Chan, Jeffrey C. Carlyle, and Roy H. Campbell. Cloaker: Hardware Supported Rootkit Concealment. In *IEEE Symposium on Security and Privacy*, pages 296–310. IEEE Computer Society, 2008.
34. Jonathan Davidson. *Voice Over IP Fundamentals*. Cisco Press Fundamentals Series. Cisco Press, 2006.
35. Guillaume Delugré. Closer to Metal: Reverse Engineering the Broadcom NetExtreme's Firmware. Sogeti ESEC Lab: http://esec-lab.sogeti.com/dotclear/public/publications/10-hack.lu-nicreverse_slides.pdf [accessed 25 February 2014], October 2010.
36. Guillaume Delugré. How to Develop a Rootkit for Broadcom NetExtreme Network Cards. Sogeti ESEC Lab: http://esec-lab.sogeti.com/dotclear/public/publications/11-reconnicreverse_slides.pdf [accessed 25 February 2014], 2011.
37. Department of Defense. DEPARTMENT OF DEFENSE TRUSTED COMPUTER SYSTEM EVALUATION CRITERIA. NIST CSRC: http://csrc.nist.gov/publications/history/dod85.pdf [accessed 25 February 2014], December 1985. DEPARTMENT OF DEFENSE STANDARD.
38. T. Dierks and E. Rescorla. The Transport Layer Security (TLS) Protocol Version 1.2. Internet Engineering Task Force: http://www.ietf.org/rfc/rfc5246.txt [accessed 25 February 2014], August 2008. Network Working Group RFC 5246.
39. Kurt Dietrich. A Secure and Reliable Platform Configuration Change Reporting Mechanism for Trusted Computing Enhanced Secure Channels. In *Proceedings of the 9th International Conference for Young Computer Scientists, 2008. ICYCS 2008*, pages 2137–2142, 2008.
40. Kurt Dietrich. On Reliable Platform Configuration Change Reporting Mechanisms for Trusted Computing Enabled Platforms. *Journal of Universal Computer Science*, 16(4):507–518, 2010.
41. Jeroen Domburg. Hard Disk Hacking. SpritesMods.com: http://spritesmods.com/?art=hddhack&page=1 [accessed 25 February 2014], 2013. Presented at OHM2013: http://bofh.nikhef.nl/events/OHM/video/d2-t1-13-20130801-2300-hard_disks_more_than_just_block_devices-sprite_tm.m4v [accessed 25 February 2014].
42. Maximilian Dornseif, Michael Becher, and Christian N. Klein. FireWire - All Your Memory Are Belong To Us. CanSecWest: http://cansecwest.com/core05/2005-firewire-cansecwest.pdf [accessed 25 February 2014], May 2005.
43. Maximillian Dornseif. 0wned by an iPod - Hacking by Firewire. Laboratory for Dependable Distributed Systems University of Mannheim: http://pi1.informatik.uni-mannheim.de/filepool/presentations/0wned-by-an-ipod-hacking-by firewire.pdf [accessed 25 February 2014], November 2004. PacSec 2004.
44. Loïc Duflot, Olivier Levillain, and Benjamin Morin. ACPI: Design Principles and Concerns. In *Proceedings of the 2nd International Conference on Trusted Computing*, Trust'09, pages 14–28, Berlin, Heidelberg, 2009. Springer-Verlag.
45. Loïc Duflot, Yves Alexis Perez, and Benjamin Morin. Run-time Firmware Integrity Verification: What If You Can't Trust Your Network Card? French Network and Information Security Agency (FNISA): http://www.ssi.gouv.fr/IMG/pdf/Duflot-Perez_runtime-firmware-integrity-verification.pdf [accessed 25 February 2014], March 2011.
46. Loïc Duflot, Yves-Alexis Perez, and Benjamin Morin. What If You Can't Trust Your Network Card? In *Proceedings of the 2011 International Symposium on Research in Attacks, Intrusions and Defenses (RAID)*, pages 378–397, 2011.
47. Loïc Duflot, Yves-Alexis Perez, Guillaume Valadon, and Olivier Levillain. Can You Still Trust Your Network Card? French Network and Information Security Agency (FNISA): http://www.ssi.gouv.fr/IMG/pdf/csw-trustnetworkcard.pdf [accessed 25 February 2014], March 2010.
48. Marcel Eckert, Igor Podebrad, and Bernd Klauer. Hardware Based Security Enhanced Direct Memory Access. In Bart Decker, Jana Dittmann, Christian Kraetzer, and Claus Vielhauer,

editors, *Communications and Multimedia Security, volume 8099 of Lecture Notes in Computer Science*, pages 145–151. Springer, Berlin Heidelberg, 2013.

49. Shawn Embleton, Sherri Sparks, and Cliff Zou. SMM Rootkits: A New Breed of OS Independent Malware. In *Proceedings of the 4th International Conference on Security and Privacy in Communication Networks*, pages 1–12, New York, NY, USA, 2008. ACM.

50. A. Freier, P. Karlton, and P. Kocher. The Secure Sockets Layer (SSL) Protocol Version 3.0. Internet Engineering Task Force: http://tools.ietf.org/html/rfc6101 [accessed 25 February 2014], August 2011. Category: Historic.

51. Tal Garfinkel and Mendel Rosenblum. A Virtual Machine Introspection Based Architecture for Intrusion Detection. In *Proceedings of the 2003 Network and Distributed Systems Security Symposium*, February 2003.

52. Yacine Gasmi, Ahmad-Reza Sadeghi, Patrick Stewin, Martin Unger, and N. Asokan. Beyond Secure Channels. In *Proceedings of the 2007 ACM Workshop on Scalable Trusted Computing*, STC'07, pages 30–40, New York, NY, USA, 2007. ACM.

53. Kenneth Goldman, Ronald Perez, and Reiner Sailer. Linking Remote Attestation to Secure Tunnel Endpoints. In *STC '06: Proceedings of the 1st ACM Workshop on Scalable Trusted Computing*, pages 21–24, New York, NY, USA, November 2006. ACM Press.

54. David Grawrock. *Dynamics of a Trusted Platform: A Building Block Approach*. Intel Press, 2009.

55. John Heasman. Implementing and Detecting a PCI Rootkit. Black Hat: http://www.blackhat.com/presentations/bh-dc-07/Heasman/Paper/bh-dc-07-Heasman-WP.pdf [accessed 25 February 2014], 2006.

56. John Heasman. Implementing and Detecting an ACPI BIOS Rootkit. Black Hat Federal: http://www.blackhat.com/presentations/bh-federal-06/BH-Fed-06-Heasman.pdf [accessed 25 February 2014], 2006.

57. John Heasman. Hacking the Extensible Firmware Interface. Black Hat USA: https://www.blackhat.com/presentations/bh-usa-07/Heasman/Presentation/bh-usa-07-heasman.pdf [accessed 25 February 2014], 2007.

58. John L. Hennessy and David A. Patterson. *Computer Architecture: A Quantitative Approach*. Morgan Kaufmann, May 2005. 3rd edition.

59. John L. Hennessy and David A. Patterson. *Computer Architecture: A Quantitative Approach*. Morgan Kaufmann, 2012. 5th edition.

60. Greg Hoglund and Jamie Butler. *Rootkits: Subverting the Windows Kernel*. Addison Wesley Professional, 2005.

61. David Hulton. Cardbus Bus-Mastering: 0Wning The Laptop, January 2006. Shmoocon 2006.

62. Intel Corporation. Universal Host Controller Interface (UHCI) Design Guide. The Slackware Linux Project: ftp://ftp.slackware.com/pub/netwinder/pub/misc/docs/29765002-usb-uhci%20design%20guide.pdf [accessed 25 February 2014], March 1996. Revision 1.1.

63. Intel Corporation. Intel 3 Series Express Chipset Family. Intel Corporation: http://www.intel.com/Assets/PDF/datasheet/316966.pdf [accessed 25 February 2014], August 2007.

64. Intel Corporation. Intel I/O Controller Hub (ICH9) Family. Intel Corporation: http://www.intel.com/content/dam/doc/datasheet/io-controller-hub-9-datasheet.pdf [accessed 25 February 2014], August 2008.

65. Intel Corporation. Intel I/O Controller Hub 8/9/10 and 82566/82567/82562V Software Developer's Manual. Intel Corporation: http://www.intel.com/content/dam/doc/manual/i-o-controller-hub-8-9-10-82566-82567-82562v-software-dev-manual.pdf [accessed 25 February 2014], July 2009.

66. Intel Corporation. 2nd Generation Intel Core vPro Processor Family. Intel Corporation: http://www.intel.com/content/dam/doc/white-paper/performance-2nd-generation-core-vpro-family-paper.pdf [accessed 25 February 2014], June 2011.

67. Intel Corporation. Access Accounts More Securely with Intel Identity Protection Technology. Intel Corporation: http://ipt.intel.com/Libraries/Documents/Intel_IdentityProtect_techbrief_v7.sflb.ashx [accessed 25 February 2014], February 2011.

68. Intel Corporation. Intel 5 Series Chipset and Intel 3400 Series Chipset. Intel Corporation: http://www.intel.com/content/dam/doc/datasheet/5-chipset-3400-chipset-datasheet. pdf [accessed 25 February 2014], January 2012.
69. Intel Corporation. Intel 64 and IA-32 Architectures Software Developer's Manual—Volume 3 (3A, 3B & 3C): System Programming Guide. Intel Corporation: http://download.intel.com/ products/processor/manual/325384.pdf [accessed 27 April 2012], March 2012.
70. Intel Corporation. Intel Architecture Instruction Set Extensions Programming Reference. Intel Corporation: http://download-software.intel.com/sites/default/files/319433-015. pdf [accessed 25 February 2014], July 2013.
71. Intel Corporation. Intel VTune Amplifier 2013 - Document Number: 326734–004. Intel Corporation: http://software.intel.com/sites/products/documentation/doclib/iss/2013/amplifier/ lin/ug_docs/index.htm [accessed 25 February 2014], 2013. External Bus Events.
72. International Business Machines Corp. IBM 4764 PCI-X Cryptographic Coprocessor. International Business Machines Corp.: http://www-03.ibm.com/security/cryptocards/pcixcc/ overview.shtml [accessed 5 March 2012], March 2012.
73. International Business Machines Corp. IBM PCIe Cryptographic Coprocessor. International Business Machines Corp.: http://www-03.ibm.com/security/cryptocards/pciecc/overview. shtml [accessed 5 March 2012], March 2012.
74. Shan Jiang, Sean Smith, and Kazuhiro Minami. Securing Web Servers against Insider Attack. In *ACSAC '01: Proceedings of the 17th Annual Computer Security Applications Conference*, page 265, Washington, DC, USA, 2001. IEEE Computer Society.
75. C. Kaufman, P. Hoffman, Y. Nir, and P. Eronen. Internet Key Exchange Protocol Version 2 (IKEv2). The Internet Engineering Task Force: http://www.ietf.org/rfc/rfc5996.txt [accessed 25 February 2014], September 2010. RFC5996.
76. S. Kent and K. Seo. Security Architecture for the Internet Protocol. Internet Engineering Task Force: http://www.ietf.org/rfc/rfc4301.txt [accessed 25 February 2014], December 2005. Network Working Group RFC 4346. Obsoletes: RCF2401.
77. Samuel T. King, Peter M. Chen, Yi-Min Wang, Chad Verbowski, Helen J. Wang, and Jacob R. Lorch. SubVirt: Implementing Malware with Virtual Machines. In *SP '06: Proceedings of the 2006 IEEE Symposium on Security and Privacy*, pages 314–327, Washington, DC, USA, 2006. IEEE Computer Society.
78. Markulf Kohlweiss, Ueli Maurer, Cristina Onete, Björn Tackmann, and Daniele Venturi. (De-)Constructing TLS. Cryptology ePrint Archive: http://eprint.iacr.org/2014/020.pdf [accessed 25 February 2014], January 2014.
79. Arvind Kumar, Purushottam Goel, and Ylian Saint-Hilaire. *Active Platform Management Demystified*. 2009. Intel Press.
80. Evangelos Ladakis, Lazaros Koromilas, Giorgos Vasiliadis, Michalis Polychronakis, and Sotiris Ioannidis. You Can Type, but You Can't Hide: A Stealthy GPU-based Keylogger. In *Proceedings of the 6th European Workshop on System Security*. EuroSec, Prague, Czech Republic, April 2013.
81. Hojoon Lee, Hyungon Moon, Daehee Jang, Kihwan Kim, Jihoon Lee, Yunheung Paek, and Brent Byunghoon Kang. KI-Mon: A Hardware-assisted Event-triggered Monitoring Platform for Mutable Kernel Object. In *Proceedings of the 22nd Conference on USENIX Security Symposium*, SSYM'13. USENIX Association, 2013.
82. Yanlin Li, Jonathan M. McCune, and Adrian Perrig. SBAP: Software-based Attestation for Peripherals. In *Proceedings of the 3rd International Conference on Trust and Trustworthy Computing*, TRUST'10, pages 16–29, Berlin, Heidelberg, 2010. Springer-Verlag.
83. Yanlin Li, Jonathan M. McCune, and Adrian Perrig. VIPER: Verifying the Integrity of PERipherals' Firmware. In *Proceedings of the ACM Conference on Computer and Communications Security (CCS)*, October 2011.
84. Loukas K (snare). DE MYSTERIIS DOM JOBSIVS Mac EFI Rootkits. ho/ax.: http://ho.ax/ downloads/De_Mysteriis_Dom_Jobsivs_Black_Hat_Paper.pdf [accessed 25 February 2014], 2012. Paper.

85. Loukas K (snare). DE MYSTERIIS DOM JOBSIVS Mac EFI Rootkits. ho/ax.: http://
ho.ax/downloads/De_Mysteriis_Dom_Jobsivs_Black_Hat_Slides.pdf [accessed 25 February
2014], 2012. Slides.

86. John Lyle and Andrew Martin. Engineering Attestable Services. In Alessandro Acquisti,
SeanW. Smith, and Ahmad-Reza Sadeghi, editors, *Trust and Trustworthy Computing, volume
6101 of Lecture Notes in Computer Science*, pages 257–264. Springer, Berlin Heidelberg,
2010.

87. Carsten Maartmann-Moe. Inception. Break & Enter: http://www.breaknenter.org/projects/
inception/ [accessed 25 February 2014].

88. Vinod Mamtani. DMA Directions And Windows. Microsoft: http://download.microsoft.com/
download/a/f/d/afdfd50d-6eb9-425e-84e1-b4085a80e34e/sys-t304_wh07.pptx   [accessed
25 February 2014], 2007.

89. John Marchesini, Sean W. Smith, Omen Wild, Josh Stabiner, and Alex Barsamian. Open-
Source Applications of TCPA Hardware. In *ACSAC '04: Proceedings of the 20th Annual
Computer Security Applications Conference (ACSAC'04)*, pages 294–303, Washington, DC,
USA, 2004. IEEE Computer Society.

90. David Maynor. DMA: Skeleton Key of Computing & & Selected Soap Box Rants.
CanSecWest: http://cansecwest.com/core05/DMA.ppt [accessed 25 February 2014], May
2005.

91. Jonathan M. McCune, Bryan Parno, Adrian Perrig, Michael K. Reiter, and Arvind Seshadri.
Minimal TCB Code Execution. In *SP '07: Proceedings of the 2007 IEEE Symposium on
Security and Privacy*, pages 267–272, Washington, DC, USA, 2007. IEEE Computer Society.

92. Hyungon Moon, Hojoon Lee, Jihoon Lee, Kihwan Kim, Yunheung Paek, and Brent Byunghoon Kang. Vigilare: Toward Snoop-based Kernel Integrity Monitor. In *Proceedings of the
2012 ACM Conference on Computer and Communications Security*, CCS'12, pages 28–37,
New York, NY, USA, 2012. ACM.

93. Tilo Müller, Andreas Dewald, and Felix C. Freiling. AESSE: A Cold-boot Resistant Implementation of AES. In *Proceedings of the Third European Workshop on System Security*,
EUROSEC '10, pages 42–47, New York, NY, USA, 2010. ACM.

94. Tilo Müller, Felix C. Freiling, and Andreas Dewald. TRESOR Runs Encryption Securely
Outside RAM. In *Proceedings of the 20th USENIX Conference on Security*, SEC'11, pages
17–17, Berkeley, CA, USA, 2011. USENIX Association.

95. Tilo Müller, Benjamin Taubmann, and Felix C. Freiling. TreVisor: OS-independent Software-
based Full Disk Encryption Secure against Main Memory Attacks. In *Proceedings of the 10th
International Conference on Applied Cryptography and Network Security*, ACNS'12, pages
66–83, Berlin, Heidelberg, 2012. Springer-Verlag.

96. Quan Nguyen. Issues in Software-based Attestation. Kaspersky Lab: http://www.kaspersky.
com/images/Quan%20Nguyen.pdf [accessed 25 February 2014], November 2012.

97. Alfredo Ortega and Anibal Sacco. Deactivate the Rootkit: Attacks on BIOS Anti-theft Tech-
nologies. Black Hat USA: http://www.blackhat.com/presentations/bh-usa-09/ORTEGA/
BHUSA09-Ortega-DeactivateRootkit-SLIDES.pdf [accessed 25 February 2014], July 2009.
Slides.

98. Alfredo Ortega and Anibal Sacco. Deactivate the Rootkit: Attacks on BIOS Anti-theft Tech-
nologies. Black Hat USA: http://www.blackhat.com/presentations/bh-usa-09/ORTEGA/
BHUSA09-Ortega-DeactivateRootkit-PAPER.pdf [accessed 25 February 2014], July 2009.
Paper.

99. Siani Pearson, Boris Balacheff, Liqun Chen, David Plaquin, and Graeme Proudler. *Trusted
Computing Platforms: TCPA Technology in Context*. Prentice Hall PTR, Upper Saddle River,
NJ, USA, 2002. Hewlett-Packard Professional Books.

100. Nick L. Petroni, Jr., Timothy Fraser, Jesus Molina, and William A. Arbaugh. Copilot - A
Coprocessor-based Kernel Runtime Integrity Monitor. In *Proceedings of the 13th Conference
on USENIX Security Symposium - Volume 13*, SSYM'04, Berkeley, CA, USA, 2004. USENIX
Association.

101. David R. Piegdon and Lexi Pimenidis. Targeting Physically Addressable Memory. In Bernhard Hämmerli and Robin Sommer, editors, *Detection of Intrusions and Malware, and Vulnerability Assessment, volume 4579 of Lecture Notes in Computer Science*, pages 193–212. Springer, Berlin Heidelberg, 2007.
102. Marsh Ray and Steve Dispensa. Renegotiating TLS. Internet Archive Way Back Machine: http://web.archive.org/web/20130203213851/http://extendedsubset.com/Renegotiating_TLS.pdf [accessed 25 February 2014], November 2009.
103. Sasha Rehbock. Trustworthy Clients: Extending TNC for Integrity Checks in Web-based Environments. Master's thesis, University of Canterbury. Computer Science and Software Engineering, 2008. http://ir.canterbury.ac.nz/handle/10092/2369 [accessed 25 February 2014].
104. James Reinders. *VTune Performance Analyzer Essentials: Measurement and Tuning Techniques for Software Developers*. Engineer to Engineer Series. Intel Press, 2005.
105. Mark E. Russinovich and David A. Solomon. *Windows Internals: Including Windows Server 2008 and Windows Vista, Fifth Edition*. Microsoft Press, 5th edition, 2009.
106. Mark E. Russinovich, David A. Solomon, and Alex Ionescu. *Windows Internals 6th Edition, Part 2*. Microsoft Press, 2012.
107. Joanna Rutkowska. Red Pill... Or How to Detect VMM Using (almost) One CPU Instruction. Internet Archive: http://web.archive.org/web/20110726182809/http://invisiblethings.org/papers/redpill.html [accessed 25 February 2014], November 2004.
108. Joanna Rutkowska. Subverting Vista Kernel for Fun and Profit. Black Hat: http://blackhat.com/presentations/bh-usa-06/BH-US-06-Rutkowska.pdf [accessed 25 February 2014], August 2006.
109. Ahmad-Reza Sadeghi and Steffen Schulz. Extending IPsec for Efficient Remote Attestation. In Radu Sion, Reza Curtmola, Sven Dietrich, Aggelos Kiayias, Josep M. Miret, Kazue Sako, and Francesc Seb, editors, *Financial Cryptography and Data Security*, volume 6054 of Lecture Notes in Computer Science, pages 150–165. Springer, Berlin Heidelberg, 2010.
110. Ahmad-Reza Sadeghi, Marko Wolf, Christian Stüble, N. Asokan, and Jan-Erik Ekberg. Enabling Fairer Digital Rights Management with Trusted Computing. In *Proceedings of the 10th International Conference on Information Security*, ISC'07, pages 53–70, Berlin, Heidelberg, 2007. Springer-Verlag.
111. Fernand Lone Sang, Éric Lacombe, Vincent Nicomette, and Yves Deswarte. Exploiting an I/OMMU Vulnerability. In *Proceedings of the 5th International Conference on Malicious and Unwanted Software (MALWARE)*, pages 7–14, October 2010.
112. Fernand Lone Sang, Vincent Nicomette, and Yves Deswarte. I/O Attacks in Intel-PC Architectures and Countermeasures. SysSec: http://www.syssec-project.eu/media/page-media/23/syssec2011-s1.4-sang.pdf [accessed 25 February 2014], July 2011.
113. S. Santesson. TLS Handshake Message for Supplemental Data. The Internet Engineering Task Force: http://www.ietf.org/rfc/rfc4680.txt [accessed 25 February 2014], September 2006. RFC4680.
114. Russ Sevinsky. Funderbolt - Adventures in Thunderbolt DMA Attacks. Black Hat: https://media.blackhat.com/us-13/US-13-Sevinsky-Funderbolt-Adventures-in-Thunderbolt-DMA-Attacks-Slides.pdf [accessed 25 February 2014], 2013.
115. Gaurav Shah, Andres Molina, and Matt Blaze. Keyboards and Covert Channels. In *Proceedings of the 15th Conference on USENIX Security Symposium - Volume 15*, USENIX-SS'06, Berkeley, CA, USA, 2006. USENIX Association.
116. Tom Shanley and Don Anderson. *ISA System Architecture*. Mindshare PC System Architecture. Addison Wesley, 1995.
117. Tom Shanley and Bob Colwell. *The Unabridged Pentium 4: IA32 Processor Genealogy*. PC System Architecture Series. Addison Wesley Professional, 2005.
118. John P. Shen and Mikko H. Lipasti. *Modern Processor Design: Fundamentals of Superscalar Processors*. Electrical and Computer Engineering. McGraw-Hill Companies, Incorporated, 2005.
119. Patrick Simmons. Security Through Amnesia: A Software-based Solution to the Cold Boot Attack on Disk Encryption. In *Proceedings of the 27th Annual Computer Security Applications Conference*, ACSAC'11, pages 73–82, New York, NY, USA, 2011. ACM.

120. Ned M. Smith. System and Method for Combining User and Platform Authentication in Negotiated Channel Security Protocols. United States Patent Application 20050216736: http://www.freepatentsonline.com/20050216736.html [accessed 25 February 2014], September 2005.

121. Stephen L. Smith. Intel Roadmap Overview August 20th 2008. Intel Corporation: http://download.intel.com/pressroom/kits/events/idffall_2008/SSmith_briefing_roadmap.pdf [accessed 25 February 2014], August 2008.

122. Patrick Stewin. A Primitive for Revealing Stealthy Peripheral-based Attacks on the Computing Platform's Main Memory. In *Proceedings of the 16th International Symposium on Research in Attacks, Intrusions and Defenses (RAID)*, 2013.

123. Patrick Stewin and Iurii Bystrov. Understanding DMA Malware. In *Proceedings of the 9th Conference on Detection of Intrusions and Malware & Vulnerability Assessment*, 2012.

124. Patrick Stewin and Iurii Bystrov. Persistent, Stealthy, Remote-controlled Dedicated Hardware Malware. http://stewin.org/slides/44con_2013-dedicated_hw_malware-stewin_bystrov.pdf [accessed 25 February 2014], September 2013. 44CON.

125. Patrick Stewin and Iurii Bystrov. Persistent, Stealthy, Remote-controlled Dedicated Hardware Malware. http://stewin.org/slides/30c3-dedicated_hw_malware-stewin_bystrov_final.pdf [accessed 25 February 2014], December 2013. 30C3: 30th Chaos Communication Congress.

126. Patrick Stewin, Jean-Pierre Seifert, and Collin Mulliner. Poster: Towards Detecting DMA Malware. In *Proceedings of the 18th ACM Conference on Computer and Communications Security*, CCS'11, pages 857–860, New York, NY, USA, 2011. ACM.

127. Jon Stokes. *Inside The Machine: An Illustrated Introduction to Microprocessors and Computer Architecture*. No Starch Press Series. No Starch Press, 2007.

128. Frederic Stumpf, Omid Tafreschi, Patrick Röder, and Claudia Eckert. A Robust Integrity Reporting Protocol for Remote Attestation. In *Proceedings of the Second Workshop on Advances in Trusted Computing (WATC'06 Fall), Tokyo*, December 2006.

129. Peter Szor. *The Art Of Computer Virus Research And Defense*. Symantec Press Series. Addison Wesley Publishing Company Incorporated, 2005.

130. TCG Infrastructure Working Group (IWG). TCG Infrastructure Working Group Reference Architecture for Interoperability (Part I). Trusted Computing Group: http://www.trustedcomputinggroup.org/files/resource_files/8770A217-1D09-3519-AD1754 3BF6163205/IWG_Architecture_v1_0_r1.pdf [accessed 25 February 2014], June 2005. Specification Version 1.0 Revision 1.

131. Alexander Tereshkin and Rafal Wojtczuk. Introducing Ring -3 Rootkits. Black Hat: http://www.blackhat.com/presentations/bh-usa-09/TERESHKIN/BHUSA09-Tereshkin-Ring3Rootkit-SLIDES.pdf [accessed 25 February 2014], July 2009.

132. The Computer Language Company Inc., Heartbeat. Computer Desktop Encyclopedia: http://lookup.computerlanguage.com/host_app/search?cid=C999999&term=heartbeat&lookup.x=27&lookup.y=21 [accessed 25 February 2014], 2013.

133. Robert Bruce Thompson and Barbara Fritchman Thompson. *PC Hardware in a Nutshell, 3rd Edition*. O'Reilly & Associates Inc, Sebastopol, CA, USA, 2003.

134. Arrigo Triulzi. Project Maux Mk.II. The Alchemist Owl: http://www.alchemistowl.org/arrigo/Papers/Arrigo-Triulzi-PACSEC08-Project-Maux-II.pdf [accessed 25 February 2014], 2008.

135. Arrigo Triulzi. The Jedi Packet Trick Takes Over the Deathstar. The Alchemist Owl: http://www.alchemistowl.org/arrigo/Papers/Arrigo-Triulzi-CANSEC10-Project-Maux-III.pdf [accessed 25 February 2014], March 2010.

136. Trusted Computing Group. TCG PC Client Specific Implementation Specification For Conventional BIOS. Trusted Computing Group: http://www.trustedcomputinggroup.org/files/temp/64505409-1D09-3519-AD5C611FAD3F799B/PC Client Implementation for BIOS.pdf [accessed 25 February 2014], July 2005.

137. Trusted Computing Group. TCG Trusted Network Connect—TNC IF-T: Binding to TLS. Trusted Computing Group: http://www.trustedcomputinggroup.org/files/static_page_files/1D8D3689-1A4B-B294-D0E7699128CB9817/TNC_IFT_TLS_v2_0_r7.pdf [accessed 25 February 2014], February 2013. Specification Version 2.0 Revision 7.

138. Trusted Network Connect Work Group. TCG Trusted Network Connect TNC Architecture for Interoperability. Trusted Computing Group: http://www.trustedcomputinggroup.org/files/resource_files/2884F884-1A4B-B294-D001FAE2E17EA3EB/TNC_Architecture_v1_5_r3-1.pdf [accessed 25 February 2014], May 2012. Specification Version 1.5, Revision 3.

139. USB Implementers Forum, Inc. USB.org - ExpressCard_specs. USB Implementers Forum Inc: http://www.usb.org/developers/expresscard/EC_specifications [accessed 25 February 2014], 2009.

140. Giorgos Vasiliadis, Michalis Polychronakis, and Sotiris Ioannidis. GPU-Assisted Malware. In *Proceedings of the 5th International Conference on Malicious and Unwanted Software (MALWARE)*, pages 1–6, October 2010.

141. Amit Vasudevan, Jonathan McCune, James Newsome, Adrian Perrig, and Leendert van Doorn. CARMA: A Hardware Tamper-resistant Isolated Execution Environment on Commodity x86 Platforms. In *Proceedings of the 7th ACM Symposium on Information, Computer and Communications Security*, ASIACCS'12, pages 48–49, New York, NY, USA, 2012. ACM.

142. Davide Vernizzi. TLS Hello Extensions and Supplemental Data. Blog: http://tlsext-general.blogspot.de/2008/12/tls-hello-extensions-and-supplemental.html [accessed 25 February 2014], December 2008.

143. Jian Wang, Zhiyong Zhang, Fei Xiang, Lili Zhang, and Qingli Chen. A Trusted Authentication Protocol based on SDIO Smart Card for DRM. *International Journal of Digital Content Technology & Its Applications*, 6(23):222–233, December 2012.

144. Filip Wecherowski. A Real SMM Rootkit: Reversing and Hooking BIOS SMI Handlers. Phrack Magazine Issue 0x42, Phile #0x0B of 0x11: http://www.phrack.org/issues.html?issue=66&id=11#article [accessed 25 February 2014], June 2009.

145. Rafal Wojtczuk and Joanna Rutkowska. Attacking SMM Memory via Intel CPU Cache Poisoning. Invisible Things Lab: http://invisiblethingslab.com/itl/Resources.html [accessed 25 February 2014], March 2009.

146. Rafal Wojtczuk and Joanna Rutkowska. Attacking Intel TXT via SINIT Code Execution Hijacking. Invisible Things Lab: http://www.invisiblethingslab.com/resources/2011/Attacking_Intel_TXT_via_SINIT_hijacking.pdf [accessed 25 February 2014], November 2011.

147. Rafal Wojtczuk and Joanna Rutkowska. Following the White Rabbit: Software Attacks against Intel VT-d Technology. Invisible Things Lab: http://www.invisiblethingslab.com/resources/2011/Software%20Attacks%20on%20Intel%20VT-d.pdf [accessed 25 February 2014], April 2011.

148. Rafal Wojtczuk, Joanna Rutkowska, and Alexander Tereshkin. Another Way to Circumvent Intel Trusted Execution Technology. Invisible Things Lab: http://invisiblethingslab.com/resources/misc09/Another%20TXT%20Attack.pdf [accessed 25 February 2014], December 2009.

149. Rafal Wojtczuk and Alexander Tereshkin. Attacking Intel BIOS. Invisible Things Lab: http://invisiblethingslab.com/resources/bh09usa/Attacking%20Intel%20BIOS.pdf [accessed 25 February 2014], July 2009.

150. Ben-Ami Yassour, Muli Ben-Yehuda, and Orit Wasserman. On the DMA Mapping Problem in Direct Device Assignment. In *Proceedings of the 3rd Annual Haifa Experimental Systems Conference*, SYSTOR'10, pages 18:1–18:12, New York, NY, USA, 2010. ACM.

151. Yue Yu, Sun Hao, and Kong Yanan. Expand the SSL/TLS Protocol on Trusted Platform Module. In *Proceedings of the International Conference on Computer Application and System Modeling (ICCASM)*, volume 11, pages V11–48-V11-51, 2010.

152. Jonas Zaddach, Anil Kurmus, Davide Balzarotti, Erik Olivier Blass, Aurelien Francillon, Travis Goodspeed, Moitrayee Gupta, and Ioannis Koltsidas. Implementation and Implications of a Stealth Hard-Drive Backdoor. In *Proceedings of the 29th Annual Computer Security Applications Conference (ACSAC)*, ACSAC 13. ACM, December 2013.

153. Fengwei Zhang. IOCheck: A Framework to Enhance the Security of I/O Devices at Runtime. In *Proceedings of the 43rd Annual IEEE/IFIP International Conference on Dependable Systems and Networks (DSN'13)*, June 2013.
154. P. Zimmermann, A. Johnston, and J. Callas. ZRTP: Media Path Key Agreement for Unicast Secure RTP. The Internet Engineering Task Force: http://www.ietf.org/rfc/rfc6189.txt [accessed 25 February 2014], April 2011. RFC6189.

Printed by Printforce, the Netherlands